WITHDRAWN

FINITE ELEMENT METHOD

THE INTEXT SERIES IN CIVIL ENGINEERING

Series Editor—**Russell C. Brinker**
  New Mexico State University

Bouchard and Moffitt—SURVEYING, 5th edition
Brinker—ELEMENTARY SURVEYING, 5th edition
Clark, Viessman, and Hammer—WATER SUPPLY AND POLLUTION CONTROL, 2nd edition
Ghali and Neville—STRUCTURAL ANALYSIS: A UNIFIED CLASSICAL AND MATRIX APPROACH
Jumikis—FOUNDATION ENGINEERING
McCormac—STRUCTURAL ANALYSIS, 2nd edition
McCormac—STRUCTURAL STEEL DESIGN, 2nd edition
Meyer—ROUTE SURVEYING AND DESIGN, 4th edition
Moffitt—PHOTOGRAMMETRY, 2nd edition
Salmon and Johnson—STEEL STRUCTURES: DESIGN AND BEHAVIOR
Spangler and Handy—SOIL ENGINEERING, 3rd edition
Ural—FINITE ELEMENT METHOD: BASIC CONCEPTS AND APPLICATIONS
Ural—MATRIX OPERATIONS AND USE OF COMPUTERS IN STRUCTURAL ENGINEERING
Viessman, Harbaugh, and Knapp—INTRODUCTION TO HYDROLOGY
Wang—COMPUTER METHODS IN ADVANCED STRUCTURAL ANALYSIS
Wang—MATRIX METHODS OF STRUCTURAL ANALYSIS
Wang and Salmon—REINFORCED CONCRETE DESIGN, 2nd edition
Winfrey—ECONOMIC ANALYSIS FOR HIGHWAYS

FINITE ELEMENT METHOD
Basic Concepts and
Applications

OKTAY URAL
University of Missouri-Rolla

Intext Educational Publishers   New York   London

Copyright © 1973 by Intext Press, Inc.

All rights reserved. No part of this book may be reprinted, reproduced, or utilized in any form or by any electronic, mechanical, or other means, now known or hereafter invented, including photocopying and recording, or in any information storage and retrieval system, without permission in writing from the Publisher.

Library of Congress Cataloging in Publication Data

Ural, Oktay.
   Finite element method.

  (The Intext series in civil engineering)
Bibliography:
  1. Structures, Theory of. 2. Finite element method. 3. Matrices. I. Title.
TA640.2.U7   624'.17   73-6838
ISBN 0-7002-2428-9

Intext Educational Publishers
257 Park Avenue South
New York, New York 10010

Text and cover design by Design 110

TA
640.2
U7

TO NURSEL, ÇİĞDEM, AND DERİN

Table of Contents

PREFACE — xi

1 MATRIX OPERATIONS RELATED TO FINITE ELEMENT METHOD — 1
Introduction — 1
Determinant of a Square Matrix — 6
Adjoint of a Matrix — 8
Matrix Inversion — 9
Comments on Matrix Operations — 28
Problems — 29
References — 31

2 MATRIX METHODS OF STRUCTURAL ANALYSIS — 33
Introduction — 33
Flexibility Matrix Method — 35
Stiffness Matrix Method — 40
Computer Program for the Analysis of Typical Structures — 63
Problems — 63
References — 65

3 THE FINITE ELEMENT METHOD IN STRUCTURAL ANALYSIS — 67
Introduction — 67
Historical Survey — 68
Theoretical Discussion — 69
Energy Principle — 71
Convergence — 72
Adaptability — 73
General Formulation of the Finite Element Method — 74
References — 77

4 PLANE STRESS AND PLANE STRAIN ANALYSIS — 79
Introduction — 79
Theoretical Development — 80
Problems — 97
References — 98

5 LONGHAND SOLUTION OF PLANE STRESS PROBLEMS — 99
Introduction — 99
Problems — 139

6 ANALYSIS OF AXISYMMETRIC SOLIDS — 141
Introduction — 141
Axisymmetric versus Plane Stress Analysis — 142
Axisymmetric Analysis — 143
Evaluation of Axisymmetric Analysis — 152
References — 153

7 ADDITIONAL ELEMENT TYPES FOR FINITE ELEMENT ANALYSIS — 155
Introduction — 155
Rectangular Plane Stress Element — 156
Triangular Element with Six Nodes — 164
Plate Bending Element — 170
Comments on the Nature of Displacement Function — 177
Comparison of the Various Elements — 181
References — 183

8 COMPUTER PROGRAMS BASED ON FINITE ELEMENT METHOD — 185
Introduction — 185
General-Purpose and Special-Purpose Programs — 186
Computer Programs — 187
ICES-STRUDL Subprogram — 188

NASTRAN Program	189
SAAS-II Program	190
ELAS Program	191
STRATA Program	191
Zienkiewicz-Cheung Program	193
Wilson Program	194
BOSOR-3 Program	194
SLADE Program	195
Program FEPLS	195
A Comparative Study	195
Problems	201
References	202

9 THE FINITE ELEMENT METHOD AND ICES-STRUDL 205
Introduction	205
ICES Programs	205
STRUDL-II Commands	208
STRUDL-II and the Finite Element Method	212
Problems	233
References	235

10 APPLICATIONS OF THE FINITE ELEMENT METHOD 237
Introduction	237
Formulation of Analysis Problems	238
Additional Applications of the Finite Element Method	257

SUPPLEMENTARY READINGS 263
I. List of Books	263
II. List of Papers	264

INDEX 269

Preface

The need for quick and accurate solutions of complex structural analysis problems and the availability of high-speed electronic digital computers paved the way for the development and applications of a comparatively new approach known as the finite element method of structural analysis. The mathematical relations associated with this procedure are based on standard matrix structural analysis methods, and matrix notations are used in the formulation of the problems, and solutions are reached by matrix operations.

By combining theory with practical results, I have tried to make the material understandable to senior students in engineering programs and practicing engineers, and I hope it will also generate interest to seek means for further applications of the method. The prerequisites are an understanding of the elementary theories of elasticity and structural analysis.

The favorable characteristics of the finite element method are the idealization of continuous complex geometric forms by a set of interconnected finite elements with known behavioral characteristics; the analysis of the elements instead of a complete structural system; the simplicity in superimposing the element solutions to develop the total results of the system; the adaptability of matrix formulation to the digital computers; and the generality of the approach for applications to the engineering fields beyond structural analysis.

Knowledge of matrix theory and familiarity with the use of digital computers are among the basic requirements of engineering education in this decade. These underlie the understanding and application of the finite element method. Its flexibility, efficiency, and generality increase the problem-solving capabilities of all engineers.

This book has six major goals:

1. To present the complete theoretical derivation of the finite element method as simply as scientifically possible using basic mathematical notation and operation.
2. To discuss the characteristics of several types of finite elements and to present an engineering evaluation of their applications.
3. To justify the theoretical derivation of the method by longhand solutions of several simple analysis problems.
4. To stress the easy and efficient applications of the method by employing several already prepared and readily available computer programs.
5. To secure a complete understanding of the method by formulating problems and obtaining solutions by the use of computer programs. Comparisons among the results obtained by different programs are tabulated to add realism.
6. To stimulate the imagination of the engineer by citing several possible areas of applications of the finite element method and presenting preliminary formulations.

The first two chapters present a background of matrix theory and matrix structural analysis methods. Chapters 3-5 present, discuss, and apply the displacement finite element method. Longhand solutions of simple plane stress problems are used to help the reader understand the method.

Chapter 6 discusses the application of the finite element method to the analysis of axisymmetric solids, using triangular torus elements. Since there are many practical engineering systems which are classified as axisymmetric, this chapter is of practical value.

Chapter 7 adds to the previously presented elements (constant strain triangle and triangular torus) four new types of elements (constant strain rectangle, linear strain triangle, rectangular plate bending element, triangular plate bending element). This chapter enlarges the scope of formulations to various types of analysis problems.

Chapter 8 introduces the reader to ten general or special purpose computer programs in order to acquaint him with the availability of various types of programs and their characteristics. Personal interests will direct him to select the ones to be considered and used for his purposes.

Chapter 9 presents the basic definitions of ICES-STRUDL, which is a general purpose program that is readily available and quickly learned. The intent here is to equip the reader with at least one program to obtain solutions to the analysis problems since the longhand solutions of complex problems by finite element formulation is neither feasible nor practical. Chapter 10 presents discussions of possible applications of the finite element method to several civil engineering problems.

I am indebted to Professors C. K. Wang, Harry West, Robert Sexsmith, and M. S. Abdulrahman for their helpful criticisms and reviews; to my own graduate students and colleagues at the University of Missouri—Rolla for their generous cooperation and discussions; to the many engineers and professors who participated in my annual short courses on the Finite Element Method for their constructive and practical comments; to Professors Logcher, Utku, Fenton, Senne, Andrews, Davis, Keith, Rang, and Avula, and Mr. Hulsey and Mr. Taner who generously helped me in these courses; to the Department of Civil Engineering, University of Missouri—Rolla for its continuous support of my work.

FINITE ELEMENT METHOD

Matrix Operations Related to Finite Element Method

INTRODUCTION

Structural engineering, which includes design and analysis problems, is becoming more dependent on digital computers. Analysis operations have been inclined toward more automation than design, due to their inherent natures. Design requires more experience and foresight, whereas analysis is relatively theoretical. This text will formulate and study the structural-analysis problems whose solutions can be used for the design calculations of the same structural system.

The use of high-speed electronic digital computers requires the matrix formulation of the structural analysis problems. This is due to the computer's capabilities to perform matrix operations rather quickly and efficiently. The engineer, who will be responsible for the definition and formulation of the analysis problem, has to be familiar with the basic matrix theory and matrix operations. Knowledge of the nature and behavior of matrices, as they are treated in this book, is more critical because the finite element method uses extensive matrix notations and operations throughout its development and applications.

Matrix concepts used by structural engineers were first introduced by Arthur Cayley [1] in 1858. It would be too ambitious to include the complete discussion of matrices in one chapter. Only the concepts and operations thought to be relevant to the finite element analysis will be considered.

A matrix is a rectangular array of quantities arranged in rows and columns which is helpful in solving a system of linear equations. The tedious algebraic operations become inefficient as the number of equations increases. A typical system of linear equations is given by the following n-expressions.

$$a_{11}x_1 + a_{12}x_2 + a_{13}x_3 + \cdots + a_{1n}x_n = b_1$$
$$a_{21}x_1 + a_{22}x_2 + a_{23}x_3 + \cdots + a_{2n}x_n = b_2$$
$$a_{31}x_1 + a_{32}x_2 + a_{33}x_3 + \cdots + a_{3n}x_n = b_3 \quad (1\text{-}1)$$
$$\vdots$$
$$a_{n1}x_1 + a_{n2}x_2 + a_{n3}x_3 + \cdots + a_{nn}x_n = b_n$$

In matrix notation, expression (1-1) is rewritten as

$$\begin{bmatrix} a_{11} & a_{12} & a_{13} & \cdots & a_{1n} \\ a_{21} & a_{22} & a_{23} & \cdots & a_{2n} \\ a_{31} & a_{32} & a_{33} & \cdots & a_{3n} \\ \vdots & & & & \\ a_{n1} & a_{n2} & a_{n3} & \cdots & a_{nn} \end{bmatrix} \begin{Bmatrix} x_1 \\ x_2 \\ x_3 \\ \\ x_n \end{Bmatrix} = \begin{Bmatrix} b_1 \\ b_2 \\ b_3 \\ \\ b_n \end{Bmatrix} \quad (1\text{-}2)$$

Expression (1-2) can be also denoted, using capital letters to represent the matrix quantities, as

$$[A][X] = [B]$$

where $[A]$ is the coefficient matrix, $[X]$ is the unknown matrix, and $[B]$ is the constant matrix.

To emphasize the value of the matrix equation, it is sufficient to rearrange it to obtain the unknowns.

$$[X] = [A]^{-1}[B] \quad (1\text{-}3)$$

A matrix is made of elements. Each element a_{ij} has a numerical value as well as position coordinates which are described by their subscripts. The first subscript defines the row number and the second subscript defines the column number of its location. An element a_{59} is at the intersection of the fifth row with the ninth column. It is convenient to use this notation because it facilitates some of the operations with matrices. A matrix having a single row is called a *row matrix*. One constituting of a single column is called a *column matrix*. The above discussion leads to the fact that matrices are rectangular arrays defined by the number of existing columns and rows. The elements of a matrix can be numbers, mathematical expressions, or alphanumerics. The matrices can be classified according to the combination of the individual values of the elements. Some matrices developed by this process have particular properties. To use them more efficiently it is helpful to be acquainted with them.

Definitions

1 NULL MATRIX: A matrix whose elements are all zero.

$$[N] = [0]$$

The product of any matrix with null matrix is a null matrix.

2 SQUARE MATRIX: A matrix having as many columns as rows. It is of a special type since it has a numerical value known as the *determinant* of the square matrix.

3 IDENTITY MATRIX: A square matrix which has all zero elements except ones at its diagonal. It is also called a *unit matrix*. The product of any compatible matrix with an identity matrix is the initial matrix; $[I][B] = [B]$.

$$[I] = \begin{bmatrix} 1 & 0 & 0 \\ 0 & 1 & 0 \\ 0 & 0 & 1 \end{bmatrix} \tag{1-4}$$

$$[A] = [I] \quad \text{if} \quad a_{ij} = 1 \quad \text{for} \quad i = j$$
$$a_{ij} = 0 \quad \text{for} \quad i \neq j$$

4 SYMMETRIC MATRIX: A matrix whose elements are symmetric with respect to the diagonal. A matrix $[A]$ is symmetric if $a_{ij} = a_{ji}$ for all $i \neq j$.

5 SKEW-SYMMETRIC MATRIX: A matrix which has a negative symmetry with respect to the diagonal. A matrix $[A]$ is skew-symmetric if $a_{ij} = -a_{ji}$ for all $i \neq j$.

6 Diagonal Matrix: A matrix whose elements are all zero except those on its diagonal. A unit matrix is a special type of a diagonal matrix. A matrix $[A]$ is a diagonal matrix if $a_{ij} = 0$ for all $i \neq j$.

7 Triangular Matrix: A matrix whose elements above or below the diagonal are all zero. There are two types of triangular matrices—*upper* and *lower*.

8 Column Matrix: A matrix having its elements arranged in a single column form.

9 Row Matrix: A matrix having its elements arranged in a single row form.

10 To Transpose a Matrix: To interchange respectively the rows and columns of a matrix. The transpose of a product of two given matrices is the same as the product of the transpose of the second by the first matrix.

$$[A \cdot B]^T = [B]^T \cdot [A]^T$$

11 To Invert a Matrix: To generate a new matrix using well-defined operations such that when it is multiplied with the original matrix it yields an identity matrix. The inversion operations are defined later in the chapter.

12 Determinant of a Matrix: A numerical value associated with each square matrix. The general expression which defines the calculation of the determinant of a matrix $[A]$ is given as

$$|A| = \sum_{i=1}^{n} a_{ij} \cdot (\text{cof } a_{ij}) = \sum_{j=1}^{n} a_{ij} \cdot (\text{cof } a_{ij})$$

The discussion of this formulation is presented by Equation (1-10).

13 Singular Matrix: A square matrix whose determinant is zero.

14 Orthogonal Matrix: A matrix whose transpose and inverse are equal. A matrix $[A]$ is an orthogonal matrix if $[A]^T = [A]^{-1}$.

Matrix Operations

In the study of structural analysis problems by the finite-element method, the familiarity with matrix operations such as addition, multiplication, inversion, and equality is necessary. Let us define these basic operations.

1 Matrix Equality: Two matrices are said to be equal if they have the same dimensions and the corresponding elements are identical.

Two matrices have the same dimensions if they have the same number of rows and columns. $[A]$ is equal to $[B]$ if (a_{ij}) is equal to (b_{ij}) for all values of the subscripts i and j.

2 MATRIX ADDITION AND SUBTRACTION: To add two or more matrices, they have to be compatible for addition. The compatibility for addition requires that the matrices have the same dimensions.

$$[A] + [B] + [C] = [D]$$
$$(a_{ij}) + (b_{ij}) + (c_{ij}) = (d_{ij})$$

The associative and commutative laws hold true for matrix addition and subtraction.

3 MULTIPLICATION OF MATRICES: To multiply two matrices the necessary condition is that the number of columns of the first must be equal to the number of rows of the second matrix.

$$[A]_{m \times n} \cdot [B]_{p \times r} = [C]_{m \times r}$$

To perform the above operation, n must be equal to p. The mathematical process of matrix multiplication is described by the following expression.

$$c_{ij} = \sum_{k=1}^{p} a_{ik} b_{kj} \tag{1-5}$$

To explain this formula, let us present an example.

$$\begin{bmatrix} 2 & 5 \\ 9 & 7 \end{bmatrix} \begin{bmatrix} 1 & 0 \\ 4 & 3 \end{bmatrix} = \begin{bmatrix} (2 \times 1) + (5 \times 4) & (2 \times 0) + (5 \times 3) \\ (9 \times 1) + (7 \times 4) & (9 \times 0) + (7 \times 3) \end{bmatrix}$$

$$= \begin{bmatrix} 2 + 20 & 0 + 15 \\ 9 + 28 & 0 + 21 \end{bmatrix} = \begin{bmatrix} 22 & 15 \\ 37 & 21 \end{bmatrix}$$

The associative and distributive laws hold true for matrix multiplication:

$$[A \cdot B] \cdot [C] = [A] \cdot [B \cdot C]$$
$$[A] \cdot [B + C] = [A] \cdot [B] + [A] \cdot [C]$$

The commutative law does not hold true for matrix multiplication:

$$[A] \cdot [B] \neq [B] \cdot [A] \tag{1-6}$$

Before discussing the matrix inversion, we need to define basic concepts which are used in inversion operation. They are the determinant and the adjoint of a matrix.

DETERMINANT OF A SQUARE MATRIX

The determinant of a square matrix is a numerical value connected with this particular matrix and computed according to a well-defined operation.

    a For a matrix $n \times n$, where n is equal or greater than 3, there exists a powerful method to determine the value of the determinant. This method is known as *expansion by cofactors*.

$$|A| = \sum_{i=1}^{n} a_{ij} \cdot (\text{cof } a_{ij}) = \sum_{j=1}^{n} a_{ij} \cdot (\text{cof } a_{ij}) \qquad (1\text{-}7)$$

The cofactor of an element a_{ij} of a matrix $[A]$ is defined as

$$(\text{cof } a_{ij}) = (-1)^{i+j} (\text{minor } a_{ij})$$

where (minor a_{ij}) is the determinant of the matrix obtained by deleting the ith row and the jth column of the given matrix. For illustration, consider the following example.

$$|A| = \begin{vmatrix} a_{11} & a_{12} & a_{13} \\ a_{21} & a_{22} & a_{23} \\ a_{31} & a_{32} & a_{33} \end{vmatrix}$$

$$= (a_{11})(-1)^{1+1} \begin{vmatrix} a_{22} & a_{23} \\ a_{32} & a_{33} \end{vmatrix} + (a_{21})(-1)^{2+1} \begin{vmatrix} a_{12} & a_{13} \\ a_{32} & a_{33} \end{vmatrix} \qquad (1\text{-}8)$$

$$+ (a_{31})(-1)^{3+1} \begin{vmatrix} a_{12} & a_{13} \\ a_{22} & a_{23} \end{vmatrix}$$

$$= a_{11}(a_{22}a_{33} - a_{23}a_{32}) - a_{21}(a_{12}a_{33} - a_{13}a_{32}) + a_{31}(a_{12}a_{23} - a_{13}a_{22})$$

    b For a 2×2 matrix, the determinant is defined as follows:

$$|A| = \begin{vmatrix} a_{11} & a_{12} \\ a_{21} & a_{22} \end{vmatrix} = a_{11}a_{22} - a_{21}a_{12} \qquad (1\text{-}9)$$

    c For a 3×3 matrix, the determinant can be computed with a cross multiplication procedure. This is good only for a 3×3 matrix.

$$|A| = \begin{vmatrix} a_{11} & a_{12} & a_{13} \\ a_{21} & a_{22} & a_{23} \\ a_{31} & a_{32} & a_{33} \end{vmatrix} = \begin{matrix} (+) & & & (-) \\ a_{11} & a_{12} & a_{13} & a_{11} & a_{12} \\ a_{21} & a_{22} & a_{23} & a_{21} & a_{22} \\ a_{31} & a_{32} & a_{33} & a_{31} & a_{32} \end{matrix} \quad (1\text{-}10)$$

$$|A| = a_{11}a_{22}a_{33} + a_{31}a_{23}a_{12} + a_{32}a_{21}a_{13} - a_{31}a_{22}a_{13}$$
$$- a_{32}a_{23}a_{11} - a_{33}a_{21}a_{12}$$

d Another effective method of evaluating a determinant with numerical elements is by the "pivotal condensation" scheme. This approach allows the computation of the determinant of a square matrix by 2 × 2 subdeterminants.

$$|A| = \begin{vmatrix} a_{11} & a_{12} & a_{13} & a_{14} & \cdots & a_{1n} \\ a_{21} & a_{22} & a_{23} & a_{24} & \cdots & a_{2n} \\ \vdots & & & & & \\ a_{n1} & a_{n2} & a_{n3} & a_{n4} & \cdots & a_{nn} \end{vmatrix}$$

$$|A| = \frac{1}{a_{11}^{n-2}} \begin{vmatrix} \begin{vmatrix} a_{11} & a_{12} \\ a_{21} & a_{22} \end{vmatrix} & \begin{vmatrix} a_{11} & a_{13} \\ a_{21} & a_{23} \end{vmatrix} & \cdots & \begin{vmatrix} a_{11} & a_{1n} \\ a_{21} & a_{2n} \end{vmatrix} \\ \vdots & & & \\ \begin{vmatrix} a_{11} & a_{12} \\ a_{n1} & a_{n2} \end{vmatrix} & \begin{vmatrix} a_{11} & a_{13} \\ a_{n1} & a_{n3} \end{vmatrix} & \cdots & \begin{vmatrix} a_{11} & a_{1n} \\ a_{n1} & a_{nn} \end{vmatrix} \end{vmatrix}$$

This approach is explained by the following numerical example.

Example 1-1

Find the determinant of the given square matrix by pivotal condensation scheme.

$$[A] = \begin{bmatrix} 1 & 0 & 4 \\ 2 & 3 & 5 \\ 0 & 1 & 1 \end{bmatrix}$$

$$|A| = \begin{vmatrix} \begin{bmatrix} 1 & 0 \\ 2 & 3 \end{bmatrix} & \begin{bmatrix} 1 & 4 \\ 2 & 5 \end{bmatrix} \\ \begin{bmatrix} 1 & 0 \\ 0 & 1 \end{bmatrix} & \begin{bmatrix} 1 & 4 \\ 0 & 1 \end{bmatrix} \end{vmatrix}$$

$$|A| = \begin{vmatrix} 3 & -3 \\ 1 & 1 \end{vmatrix}$$

$$|A| = 3 + 3 = 6$$

To facilitate the calculation of determinants, the following six rules will hold true:

- **a** Interchanging two rows or two columns of a square matrix will alter only the sign of its determinant.
- **b** Multiplying each element of a row or a column of a square matrix by a constant is the same as multiplying the determinant of the matrix by the same constant.
- **c** If two rows or columns of a square matrix are equal or multiple of each other, the value of the determinant is zero.
- **d** If all the elements of a row or a column are zero, then the determinant is zero.
- **e** The transpose of a square matrix will not change its determinant.
- **f** The determinant of a square matrix will not change if a column or a row is multiplied by a constant and added to a column or a row.

ADJOINT OF A MATRIX

The adjoint of a matrix $[A]$ is a matrix which is the transpose of the cofactor matrix of $[A]$. In general notation, this is expressed as

$$\text{adj}\,[A] = [\text{cof}\,A]^T \qquad (1\text{-}11)$$

Example 1-2

Determine the adjoint of a matrix in general notation. The given matrix $[A]$ is shown below.

$$[A] = \begin{bmatrix} a_{11} & a_{12} & a_{13} \\ a_{21} & a_{22} & a_{23} \\ a_{31} & a_{32} & a_{33} \end{bmatrix}$$

Step 1. Find the cofactor matrix of A.

$$\text{cof}\,[A] = (-1)^{i+j}\,(\text{minor of }[A])$$

$$\text{cof}[a_{ij}] = \begin{bmatrix} \text{cof } a_{11} & \text{cof } a_{12} & \text{cof } a_{13} \\ \text{cof } a_{21} & \text{cof } a_{22} & \text{cof } a_{23} \\ \text{cof } a_{31} & \text{cof } a_{32} & \text{cof } a_{33} \end{bmatrix}$$

$$= \begin{bmatrix} (a_{22}a_{33} - a_{23}a_{32}) & -(a_{21}a_{33} - a_{23}a_{31}) & (a_{21}a_{32} - a_{22}a_{31}) \\ -(a_{12}a_{33} - a_{13}a_{32}) & (a_{11}a_{33} - a_{13}a_{31}) & -(a_{11}a_{32} - a_{12}a_{31}) \\ (a_{12}a_{23} - a_{13}a_{22}) & -(a_{11}a_{23} - a_{13}a_{21}) & (a_{11}a_{22} - a_{21}a_{12}) \end{bmatrix}$$

Step 2. Transpose the cofactor matrix:

$$\text{adj}[A] = [\text{cof } A]^T = \begin{bmatrix} \text{cof } a_{11} & \text{cof } a_{21} & \text{cof } a_{31} \\ \text{cof } a_{12} & \text{cof } a_{22} & \text{cof } a_{32} \\ \text{cof } a_{13} & \text{cof } a_{23} & \text{cof } a_{33} \end{bmatrix}$$

MATRIX INVERSION

There are several methods available for matrix inversion. We will explain and illustrate the relevant methods. One has to remember that only square matrices have inverses.

1. Inversion by Adjoint Method

$$[A]^{-1} = \frac{\text{adj}[A]}{|A|} \tag{1-12}$$

Example 1-3
Solve the given system of linear equations.

$$2x + 8y + 4z = 1$$
$$7x + 3y + 5z = 2$$
$$x \quad\quad\quad + 6z = 3$$

In matrix notation we can write

$$\begin{bmatrix} 2 & 8 & 4 \\ 7 & 3 & 5 \\ 1 & 0 & 6 \end{bmatrix} \begin{Bmatrix} x \\ y \\ z \end{Bmatrix} = \begin{Bmatrix} 1 \\ 2 \\ 3 \end{Bmatrix}$$

Then we need to solve the equation for the unknowns:

$$\begin{Bmatrix} x \\ y \\ z \end{Bmatrix} = \begin{bmatrix} 2 & 8 & 4 \\ 7 & 3 & 5 \\ 1 & 0 & 6 \end{bmatrix}^{-1} \begin{Bmatrix} 1 \\ 2 \\ 3 \end{Bmatrix}$$

a The determinant of the coefficient matrix, by expression (1-10).

$$|A| = 2[(3 \times 6) - (5 \times 0)] - 7[(8 \times 6) - (4 \times 0)]$$
$$+ 1[(8 \times 5) - (4 \times 3)]$$
$$|A| = 36 - 336 + 28 = -272$$

b
$$\text{Cofactor } [A] = \begin{bmatrix} 18 & -37 & -3 \\ -48 & 8 & 8 \\ 28 & 18 & -50 \end{bmatrix}$$

c
$$\text{Adjoint of } [A] = \begin{bmatrix} 18 & -48 & 28 \\ -37 & 8 & 18 \\ -3 & 8 & -50 \end{bmatrix}$$

d Inverse of $[A] = \dfrac{\text{adj } [A]}{|A|}$ which is given by expression (1-12).

$$[A]^{-1} = -\frac{1}{272} \begin{bmatrix} 18 & -48 & 28 \\ -37 & 8 & 18 \\ -3 & 8 & -50 \end{bmatrix}$$

e The solution of the system is

$$\begin{Bmatrix} x \\ y \\ z \end{Bmatrix} = -\frac{1}{272} \begin{bmatrix} 18 & -48 & 28 \\ -37 & 8 & 18 \\ -3 & 8 & -50 \end{bmatrix} \begin{Bmatrix} 1 \\ 2 \\ 3 \end{Bmatrix} = \frac{1}{272} \begin{Bmatrix} -6 \\ -33 \\ 137 \end{Bmatrix}$$

Example 1-4

Solve the given system of linear equations. Use the adjoint method for the necessary matrix-inversion operation.

$$-2x + y + 3z = 1$$
$$4x - z = 2$$
$$3x + 3y + 2z = 3$$

Solution
The above system can be written in general matrix notation as expression (1-3):

$$[A]\{X\} = \{B\}$$

which can be rearranged to solve for the unknown matrix $\{X\}$:

$$\{X\} = [A]^{-1}\{B\}$$

The coefficient matrix $[A]$ is

$$[A] = \begin{bmatrix} -2 & 1 & 3 \\ 4 & 0 & -1 \\ 3 & 3 & 2 \end{bmatrix}$$

The determinant of the matrix $[A]$ is computed by the procedure defined by expression (1-10):

$$|A| = -2(3) - 1(8 + 3) + 3(12) = 19$$

The inverse of the matrix $[A]$ is obtained by the procedure (1-12):

$$\text{adj}\,[A] = \begin{bmatrix} 3 & 7 & -1 \\ -11 & -13 & 10 \\ 12 & 9 & -4 \end{bmatrix}$$

The inverse of $[A]$ is therefore

$$[A]^{-1} = \frac{1}{19} \begin{bmatrix} 3 & 7 & -1 \\ -11 & -13 & 10 \\ 12 & 9 & -4 \end{bmatrix}$$

The unknown matrix can be computed with a single matrix multiplication:

$$\begin{Bmatrix} x \\ y \\ z \end{Bmatrix} = \frac{1}{19} \begin{bmatrix} 3 & 7 & -1 \\ -11 & -13 & 10 \\ 12 & 9 & -4 \end{bmatrix} \begin{Bmatrix} 1 \\ 2 \\ 3 \end{Bmatrix}; \quad \begin{Bmatrix} x \\ y \\ z \end{Bmatrix} = \frac{1}{19} \begin{Bmatrix} 14 \\ -7 \\ 18 \end{Bmatrix}$$

Example 1-5
Show that the inverse of the matrix $[A]$ obtained in Example 1-4 is correct.

Solution

The product of a matrix by its inverse yields an identity matrix. If this condition is satisfied, the inverse is the correct one. Mathematically this condition is written as follows:

$$[A] \cdot [A]^{-1} = [I]$$

Now let us perform the necessary operations;

$$[A] \cdot [A]^{-1} = \begin{bmatrix} -2 & 1 & 3 \\ 4 & 0 & -1 \\ 3 & 3 & 2 \end{bmatrix} \begin{bmatrix} \frac{3}{19} & \frac{7}{19} & -\frac{1}{19} \\ -\frac{11}{19} & -\frac{13}{19} & \frac{10}{19} \\ \frac{12}{19} & \frac{9}{19} & -\frac{4}{19} \end{bmatrix}$$

This can be expanded in the form

$$= \frac{1}{19} \begin{bmatrix} (-6 - 11 + 36) & (-14 - 13 + 27) & (2 + 10 - 12) \\ (12 + 0 - 12) & (28 + 0 - 9) & (-4 + 0 + 4) \\ (9 - 33 + 24) & (21 - 39 + 18) & (-3 + 30 - 8) \end{bmatrix}$$

This expression is algebraically reduced to a simple matrix:

$$[A] \cdot [A]^{-1} = \begin{bmatrix} 1 & 0 & 0 \\ 0 & 1 & 0 \\ 0 & 0 & 1 \end{bmatrix}$$

which is the identity matrix. Hence $[A]^{-1}$ is the correct inverse of the coefficient matrix $[A]$.

2. Gauss-Jordan Method

This method is an elimination procedure without decreasing the number of equations in the system. A unit matrix of the same dimension as the original matrix is attached to the right of the original matrix. Performing the proper matrix row operations, the original matrix is reduced to a unit matrix; the same row operations are applied to the attached unit matrix, which becomes the inverse matrix. $[(A \mid I)]$ will become $[(I \mid B)]$ when $[B] = [A]^{-1}$:

$$\begin{bmatrix} a_{11} & a_{12} & a_{13} & | & 1 & 0 & 0 \\ a_{21} & a_{22} & a_{23} & | & 0 & 1 & 0 \\ a_{31} & a_{32} & a_{33} & | & 0 & 0 & 1 \end{bmatrix} \rightarrow \begin{bmatrix} 1 & 0 & 0 & | & b_{11} & b_{12} & b_{13} \\ 0 & 1 & 0 & | & b_{21} & b_{22} & b_{23} \\ 0 & 0 & 1 & | & b_{31} & b_{32} & b_{33} \end{bmatrix} \quad (1\text{-}13)$$

Example

Solve the given system by Gauss-Jordan method.

$$[A][X] = [C]$$

If $[B] = [A]^{-1}$,

$$[X] = [B][C]$$

The given system of equations is

$$x + 0y + 4z + 7w = 1$$
$$2x + 3y + 5z + 8w = 4$$
$$0x + y + z + 4w = 3$$
$$5x + 4y + z + 0w = 0$$

which in matrix notation is written as

$$\begin{bmatrix} 1 & 0 & 4 & 7 \\ 2 & 3 & 5 & 8 \\ 0 & 1 & 1 & 4 \\ 5 & 4 & 1 & 0 \end{bmatrix} \begin{Bmatrix} x \\ y \\ z \\ w \end{Bmatrix} = \begin{Bmatrix} 1 \\ 4 \\ 3 \\ 0 \end{Bmatrix}$$

To calculate the variables, the matrix transformations are performed:

$$\begin{Bmatrix} x \\ y \\ z \\ w \end{Bmatrix} = \begin{bmatrix} 1 & 0 & 4 & 7 \\ 2 & 3 & 5 & 8 \\ 0 & 1 & 1 & 4 \\ 5 & 4 & 1 & 0 \end{bmatrix}^{-1} \begin{Bmatrix} 1 \\ 4 \\ 3 \\ 0 \end{Bmatrix}$$

The Gauss-Jordan method intends to transform the coefficient matrix into an identity matrix; meantime the attached identity matrix will be the inverse matrix. To illustrate the procedure few steps of operations are explained.

a Divide the first row by the leading coefficient.
b Multiply the first row by the leading coefficient of the second row and then subtract it from the second row.
c Repeat the same process for third row, etc.

These three operations and other similar operations will develop the following matrix quantities.

14   MATRIX OPERATIONS RELATED TO FINITE ELEMENT METHOD

$$\begin{vmatrix} 1 & 0 & 4 & 7 & | & 1 & 0 & 0 & 0 \\ 2 & 3 & 5 & 8 & | & 0 & 1 & 0 & 0 \\ 0 & 1 & 1 & 4 & | & 0 & 0 & 1 & 0 \\ 5 & 4 & 1 & 0 & | & 0 & 0 & 0 & 1 \end{vmatrix}$$

$$\begin{vmatrix} 1 & 0 & 4 & 7 & | & 1 & 0 & 0 & 0 \\ 0 & 3 & -3 & -6 & | & -2 & 1 & 0 & 0 \\ 0 & 1 & 1 & 4 & | & 0 & 0 & 1 & 0 \\ 0 & 4 & -19 & -35 & | & -5 & 0 & 0 & 1 \end{vmatrix}$$

$$\begin{vmatrix} 1 & 0 & 4 & 7 & | & 1 & 0 & 0 & 0 \\ 0 & 1 & -1 & -2 & | & -\frac{2}{3} & \frac{1}{3} & 0 & 0 \\ 0 & 0 & 2 & 6 & | & +\frac{2}{3} & -\frac{1}{3} & 1 & 0 \\ 0 & 0 & -15 & -27 & | & -\frac{7}{3} & -\frac{4}{3} & 0 & 1 \end{vmatrix}$$

$$\begin{vmatrix} 1 & 0 & 0 & -5 & | & -\frac{1}{3} & \frac{2}{3} & -2 & 0 \\ 0 & 1 & 0 & 1 & | & -\frac{1}{3} & \frac{1}{6} & \frac{1}{2} & 0 \\ 0 & 0 & 1 & 3 & | & \frac{1}{3} & -\frac{1}{6} & \frac{1}{2} & 0 \\ 0 & 0 & 0 & 18 & | & \frac{8}{3} & -\frac{23}{6} & \frac{15}{2} & 1 \end{vmatrix}$$

$$\begin{vmatrix} 1 & 0 & 0 & 0 & | & 0.408 & -0.378 & -0.084 & 0.314 \\ 0 & 1 & 0 & 0 & | & -0.480 & 0.376 & 0.083 & -0.055 \\ 0 & 0 & 1 & 0 & | & -0.111 & 0.472 & -0.750 & -0.167 \\ 0 & 0 & 0 & 1 & | & 0.147 & -0.213 & 0.416 & 0.555 \end{vmatrix}$$

Then we can write the following matrix equation:

$$\begin{Bmatrix} x \\ y \\ z \\ w \end{Bmatrix} = \begin{bmatrix} 0.408 & -0.398 & -0.083 & 0.280 \\ -0.481 & 0.380 & 0.084 & -0.055 \\ -0.111 & 0.472 & -0.750 & -0.167 \\ 0.148 & -0.213 & 0.416 & 0.055 \end{bmatrix} \begin{Bmatrix} 1 \\ 4 \\ 3 \\ 0 \end{Bmatrix}$$

$$\begin{Bmatrix} x \\ y \\ z \\ w \end{Bmatrix} = \begin{Bmatrix} -0.935 \\ 1.291 \\ -0.473 \\ 0.544 \end{Bmatrix}$$

Example 1-6

Compute the inverse of the given square matrix by the Gauss-Jordan method.

$$[A] = \begin{bmatrix} 2 & 8 & 4 \\ 7 & 3 & 5 \\ 1 & 0 & 6 \end{bmatrix}$$

Gauss-Jordan operations are shown below,

$$\begin{vmatrix} 2 & 8 & 4 & | & 1 & 0 & 0 \\ 7 & 3 & 5 & | & 0 & 1 & 0 \\ 1 & 0 & 6 & | & 0 & 0 & 1 \end{vmatrix}$$

$$\begin{vmatrix} 1 & 8 & -2 & | & 1 & 0 & -1 \\ 7 & 3 & 5 & | & 0 & 1 & 0 \\ 1 & 0 & 6 & | & 0 & 0 & 1 \end{vmatrix}$$

$$\begin{vmatrix} 1 & 8 & -2 & | & 1 & 0 & -1 \\ 0 & -53 & 19 & | & -7 & 1 & 7 \\ 0 & -8 & 8 & | & -1 & 0 & 2 \end{vmatrix}$$

$$\begin{vmatrix} 1 & 8 & -2 & | & 1 & 0 & -1 \\ 0 & 1 & -\frac{19}{53} & | & \frac{7}{53} & -\frac{1}{53} & -\frac{7}{53} \\ 0 & 1 & -1 & | & \frac{1}{8} & 0 & -\frac{1}{4} \end{vmatrix}$$

$$\begin{vmatrix} 1 & 8 & -2 & 1 & 0 & -1 \\ 0 & 1 & -\dfrac{19}{53} & \dfrac{7}{53} & -\dfrac{1}{53} & -\dfrac{7}{53} \\ 0 & 0 & -\dfrac{34}{53} & -\dfrac{3}{424} & \dfrac{1}{53} & -\dfrac{25}{212} \end{vmatrix}$$

$$\begin{vmatrix} 1 & 8 & -2 & 1 & 0 & -1 \\ 0 & 1 & -\dfrac{19}{53} & \dfrac{7}{53} & -\dfrac{1}{53} & -\dfrac{7}{53} \\ 0 & 0 & 1 & \dfrac{3}{272} & -\dfrac{1}{34} & \dfrac{25}{136} \end{vmatrix}$$

$$\begin{vmatrix} 1 & 8 & -2 & 1 & 0 & -1 \\ 0 & 1 & 0 & \dfrac{37}{272} & -\dfrac{1}{34} & -\dfrac{9}{136} \\ 0 & 0 & 1 & \dfrac{3}{272} & -\dfrac{1}{34} & \dfrac{25}{136} \end{vmatrix}$$

$$\begin{vmatrix} 1 & 0 & 0 & -\dfrac{18}{272} & \dfrac{6}{34} & -\dfrac{14}{136} \\ 0 & 1 & 0 & \dfrac{37}{272} & -\dfrac{1}{34} & -\dfrac{9}{136} \\ 0 & 0 & 1 & \dfrac{3}{272} & -\dfrac{1}{34} & \dfrac{25}{136} \end{vmatrix}$$

Therefore the inverse of the given matrix $[A]$ is

$$[A]^{-1} = \begin{bmatrix} -\dfrac{18}{272} & \dfrac{6}{34} & -\dfrac{14}{136} \\ \dfrac{37}{272} & -\dfrac{1}{34} & -\dfrac{9}{136} \\ \dfrac{3}{272} & -\dfrac{1}{34} & \dfrac{25}{136} \end{bmatrix}$$

3. Inversion by Partitioning

The definition of a matrix does not rule out the possibility of the elements of a matrix themselves being matrices. Usually it is helpful to partition a matrix into submatrices. For the product of two matrices it is convenient to partition them before performing the multiplication. The matrices for which this is true are said to be *conformally partitioned*. The partitions have to have well-defined correspondence among themselves for the requirement of compatibility.

$$[A] = \begin{bmatrix} A_{11} & A_{12} \\ A_{21} & A_{22} \end{bmatrix} = \begin{bmatrix} a_{11} & a_{12} & a_{13} \\ a_{21} & a_{22} & a_{23} \\ a_{31} & a_{32} & a_{33} \end{bmatrix}$$

Let $[B] = [b_{ij}]$ be the inverse of $[A] = [a_{ij}]$. Then we can write the following matrix equations:

$$[A][B] = [I]$$
$$[a_{ij}][b_{ij}] = [I]$$

Let us proceed with the following operations to illustrate the inversion by partitioning.

$$\begin{bmatrix} A_{11} & A_{12} \\ A_{21} & A_{22} \end{bmatrix} \begin{bmatrix} B_{11} & B_{12} \\ B_{21} & B_{22} \end{bmatrix} = \begin{bmatrix} I & 0 \\ 0 & I \end{bmatrix}$$

Let us expand the above matrix expression to obtain the needed relevant formulation. According to one-to-one correspondence in the equality of matrices, we will have the following expressions:

$$A_{11} B_{11} + A_{12} B_{21} = I \tag{1-14}$$
$$A_{11} B_{12} + A_{12} B_{22} = 0 \tag{1-15}$$
$$A_{21} B_{11} + A_{22} B_{21} = 0 \tag{1-16}$$
$$A_{21} B_{12} + A_{22} B_{22} = I \tag{1-17}$$

By algebraic manipulations, one can express all B terms as functions of A terms.

$$B = \phi(A)$$

from expression (1-15), B_{12} is calculated by

$$B_{12} = A_{11}^{-1}(-A_{12} B_{22}) \tag{1-18}$$

Then substitute Equation (1-18) in Equation (1-17) and solve for B_{22}:

$$B_{22} = (A_{22} - A_{21} A_{11}^{-1} A_{12})^{-1} \tag{1-19}$$

From expression (1-14), find B_{11} by

$$B_{11} = A_{11}^{-1}(I - A_{12} B_{21}) \tag{1-20}$$

Now substitute (1-20) in expression (1-16) to obtain B_{21}:

$$B_{21} = (A_{22} - A_{21}A_{11}^{-1}A_{12})^{-1}(-A_{21}A_{11}^{-1})$$
$$B_{21} = B_{22}(-A_{21}A_{11}^{-1}) \quad (1\text{-}21)$$

Hence this approach will allow the determination of the inverse of a given square matrix by actually computing only the inverse of (A_{11}).

Example 1-7

Solve the given system of linear equations by the use of inversion of the coefficient matrix by the method of partitioning.

$$2x + 8y + 4z = 1$$
$$7x + 3y + 5z = 2$$
$$x + 6z = 3$$

Solution

The above system can be written in matrix form:

$$\begin{bmatrix} 2 & 8 & 4 \\ 7 & 3 & 5 \\ 1 & 0 & 6 \end{bmatrix} \begin{Bmatrix} x \\ y \\ z \end{Bmatrix} = \begin{Bmatrix} 1 \\ 2 \\ 3 \end{Bmatrix}$$

Step 1. The inversion of a 1×1 matrix, $\{2\}$:

$$(A_{11})^{-1} = \tfrac{1}{2}$$

Step 2. The inversion of a 2×2 matrix:

$$\begin{bmatrix} 2 & 8 \\ 7 & 3 \end{bmatrix} = \begin{bmatrix} A_{11} & A_{12} \\ A_{21} & A_{22} \end{bmatrix}$$

(B_{22}) is computed according to expression (1-19):

$$B_{22} = \frac{1}{3 - (7 \cdot \tfrac{1}{2} \cdot 8)} = -\frac{1}{25}$$

(B_{12}) is computed according to expression (1-18):

$$B_{12} = \left(-\tfrac{1}{2}\right)(8)\left(-\tfrac{1}{25}\right) = \tfrac{4}{25}$$

(B_{21}) is computed according to expression (1-21):

$$B_{21} = -\left(-\tfrac{1}{25}\right)(7)\left(\tfrac{1}{2}\right) = \tfrac{7}{50}$$

(B_{11}) is computed according to expression (1-20):

$$B_{11} = \frac{1}{2} - \left(\frac{1}{2} \cdot 8 \cdot \frac{7}{50}\right) = -\frac{3}{50}$$

Then the inverse of the two by two matrix will be

$$\begin{bmatrix} 2 & 8 \\ 7 & 3 \end{bmatrix}^{-1} = \begin{bmatrix} -\frac{3}{50} & \frac{4}{25} \\ \frac{7}{50} & -\frac{1}{25} \end{bmatrix}$$

Step 3. The inversion of the 3 × 3 matrix will be obtained using the same scheme as illustrated above.

$$[A] = \begin{bmatrix} 2 & 8 & | & 4 \\ 7 & 3 & | & 5 \\ -- & -- & + & -- \\ 1 & 0 & | & 6 \end{bmatrix} = \begin{bmatrix} A_{11} & | & A_{12} \\ -- & + & -- \\ A_{21} & | & A_{22} \end{bmatrix}$$

$$A_{11}^{-1} A_{12} = \begin{bmatrix} -\frac{3}{50} & \frac{4}{25} \\ \frac{7}{50} & -\frac{1}{25} \end{bmatrix} \begin{bmatrix} 4 \\ 5 \end{bmatrix} = \begin{bmatrix} \frac{14}{25} \\ \frac{18}{50} \end{bmatrix}$$

$$A_{21} A_{11}^{-1} A_{12} = (1 \quad 0) \begin{bmatrix} \frac{14}{25} \\ \frac{18}{50} \end{bmatrix} = \begin{bmatrix} \frac{14}{25} \end{bmatrix}$$

$$B_{22} = (A_{22} - A_{21} A_{11}^{-1} A_{12})^{-1} = \begin{bmatrix} 6 - \frac{14}{25} \end{bmatrix}^{-1} = \begin{bmatrix} \frac{25}{136} \end{bmatrix}$$

$$B_{12} = -A_{11}^{-1} A_{12} B_{22} = -\begin{bmatrix} \frac{14}{25} \\ \frac{18}{50} \end{bmatrix} \begin{bmatrix} \frac{25}{136} \end{bmatrix} = -1 \cdot \begin{bmatrix} \frac{14}{136} \\ \frac{18}{272} \end{bmatrix}$$

$$B_{21} = A_{21} A_{11}^{-1} \cdot [-B_{22}] = \left(-\frac{25}{136}\right)(1 \quad 0) \begin{bmatrix} -\frac{3}{50} & \frac{4}{25} \\ \frac{7}{50} & -\frac{1}{25} \end{bmatrix}$$

$$= \begin{bmatrix} \dfrac{3}{272} & -\dfrac{4}{136} \end{bmatrix}$$

$$B_{11} = A_{11}^{-1} = A_{11}^{-1} \cdot A_{12} B_{21} = \begin{bmatrix} -\dfrac{3}{50} & \dfrac{4}{25} \\ \dfrac{7}{50} & -\dfrac{1}{25} \end{bmatrix} - \begin{bmatrix} \dfrac{14}{25} \\ \dfrac{18}{50} \end{bmatrix}$$

$$\begin{bmatrix} \dfrac{3}{272} & -\dfrac{4}{136} \end{bmatrix} = \begin{bmatrix} -\dfrac{18}{272} & \dfrac{24}{136} \\ \dfrac{37}{272} & -\dfrac{4}{136} \end{bmatrix}$$

$$[A]^{-1} = \begin{bmatrix} B_{11} & B_{12} \\ B_{21} & B_{22} \end{bmatrix} = \begin{bmatrix} -\dfrac{18}{272} & \dfrac{24}{136} & -\dfrac{14}{136} \\ \dfrac{37}{272} & -\dfrac{4}{136} & -\dfrac{18}{272} \\ \dfrac{3}{272} & -\dfrac{4}{136} & \dfrac{25}{136} \end{bmatrix}$$

$$[A]^{-1} = \dfrac{1}{272} \cdot \begin{bmatrix} -18 & 48 & -28 \\ 37 & -8 & -18 \\ 3 & -8 & 50 \end{bmatrix}$$

Once the inverse of $[A]$ is obtained, the unknown matrix can be readily computed by a mere matrix multiplication:

$$\begin{Bmatrix} x \\ y \\ z \end{Bmatrix} = \dfrac{1}{272} \begin{bmatrix} -18 & 48 & -28 \\ 37 & -8 & -18 \\ 3 & -8 & 50 \end{bmatrix} \begin{Bmatrix} 1 \\ 2 \\ 3 \end{Bmatrix}$$

$$\begin{Bmatrix} x \\ y \\ z \end{Bmatrix} = \dfrac{1}{272} \begin{Bmatrix} -6 \\ -33 \\ 137 \end{Bmatrix}$$

Example 1-8

Perform the multiplication of the two given partitioned matrices.

$$\{A\} = \left\{\begin{array}{cc|c} 1 & 2 & -1 \\ \hline 3 & 0 & 2 \end{array}\right\}$$

$$\{B\} = \left\{\begin{array}{cc} 3 & 1 \\ 1 & 3 \\ \hline 2 & 0 \end{array}\right\}$$

Solution

$$\{A\} \cdot \{B\} = \left\{\begin{array}{cc|c} 1 & 2 & -1 \\ \hline 3 & 0 & 2 \end{array}\right\} \cdot \left\{\begin{array}{cc} 3 & 1 \\ 1 & 3 \\ \hline 2 & 0 \end{array}\right\}$$

Let $(A_{11}) = \begin{pmatrix} 1 & 2 \end{pmatrix}$
$(A_{12}) = (-1)$
$(A_{21}) = \begin{pmatrix} 3 & 0 \end{pmatrix}$
$(A_{22}) = (2)$
$(B_{11}) = \begin{bmatrix} 3 & 1 \\ 1 & 3 \end{bmatrix}$
$(B_{21}) = \begin{pmatrix} 2 & 0 \end{pmatrix}$

Then the following multiplication operations are performed:

$$\begin{bmatrix} A_{11} & A_{12} \\ \hline A_{21} & A_{22} \end{bmatrix} \begin{bmatrix} B_{11} \\ B_{21} \end{bmatrix} = \begin{bmatrix} A_{11} B_{11} + A_{12} B_{21} \\ A_{21} B_{11} + A_{22} B_{21} \end{bmatrix}$$

$$= \begin{bmatrix} (1 \quad 2) \begin{bmatrix} 3 & 1 \\ 1 & 3 \end{bmatrix} + (-1)(2 \quad 0) \\ (3 \quad 0) \begin{bmatrix} 3 & 1 \\ 1 & 3 \end{bmatrix} + (2)(2 \quad 0) \end{bmatrix}$$

$$= \begin{bmatrix} (5 \quad 7) + (-2 \quad 0) \\ (9 \quad 3) + (4 \quad 0) \end{bmatrix}$$

This will give the result of the multiplication:

$$\{A\} \cdot \{B\} = \begin{bmatrix} 3 & 7 \\ 13 & 3 \end{bmatrix}$$

Cholesky's Method

This scheme for the solution of a system of linear equations related to structural analysis is very desirable. Theoretically, Cholesky's method is based on the fact that any square matrix can be expressed as the product of an upper-triangular matrix and a lower-triangular matrix. This concept was introduced in the United States by P. D. Crout [2] in 1941.

Let $[A]$ be a square matrix, $[T]$ an upper triangular matrix, and $[L]$ a lower triangular matrix such that they are related by the following general matrix equation:

$$[A] = [L][T] \quad (1\text{-}22)$$

If the square matrix $[A]$ is a 4×4 matrix, then one can write

$$\begin{bmatrix} A_{11} & A_{12} & A_{13} & A_{14} \\ A_{21} & A_{22} & A_{23} & A_{24} \\ A_{31} & A_{32} & A_{33} & A_{34} \\ A_{41} & A_{42} & A_{43} & A_{44} \end{bmatrix} = \begin{bmatrix} 1 & 0 & 0 & 0 \\ L_{21} & 1 & 0 & 0 \\ L_{31} & L_{32} & 1 & 0 \\ L_{41} & L_{42} & L_{43} & 1 \end{bmatrix} \cdot \begin{bmatrix} T_{11} & T_{12} & T_{13} & T_{14} \\ 0 & T_{22} & T_{23} & T_{24} \\ 0 & 0 & T_{33} & T_{34} \\ 0 & 0 & 0 & T_{44} \end{bmatrix}$$

$$(1\text{-}23)$$

The expansion of the above matrix quantities (Equation 1-23) will yield the values of the elements of matrices $[L]$ and $[T]$ as functions of the elements of the matrix $[A]$. To illustrate the procedure, few steps of the multiplication are performed.

$$A_{11} = T_{11}$$

$$A_{21} = L_{21}T_{11}; \quad L_{21} = \frac{A_{21}}{T_{11}} = \frac{A_{21}}{A_{11}}$$

$$A_{31} = L_{31}T_{11}; \quad L_{31} = \frac{A_{31}}{T_{11}} = \frac{A_{31}}{A_{11}}$$

$$A_{41} = L_{41}T_{11}; \quad L_{41} = \frac{A_{41}}{T_{11}} = \frac{A_{41}}{A_{11}}$$

$$A_{12} = T_{12}$$

$$A_{22} = L_{21}T_{12} + T_{22}; \quad T_{22} = A_{22} - L_{21}T_{12} = A_{22} - \frac{A_{21}}{A_{11}} \cdot A_{12}$$

$$A_{32} = L_{31}T_{12} + L_{32}T_{22}; \quad L_{32} = (A_{32} - L_{31}T_{12})/T_{22}$$
$$A_{42} = L_{41}T_{12} + L_{42}T_{22}; \quad L_{42} = (A_{42} - L_{41}T_{12})/T_{22}$$

Once all the elements of the upper and lower triangular matrices are determined, the inversion process becomes a comparatively simple operation. To clarify this, let us invert the upper triangular matrix. A matrix $[S]$ is assumed to be the searched inverse. Then Equation (1-24) is written:

$$\begin{bmatrix} S_{11} & S_{12} & S_{13} & S_{14} \\ 0 & S_{22} & S_{23} & S_{24} \\ 0 & 0 & S_{33} & S_{34} \\ 0 & 0 & 0 & S_{44} \end{bmatrix} \cdot \begin{bmatrix} T_{11} & T_{12} & T_{13} & T_{14} \\ 0 & T_{22} & T_{23} & T_{24} \\ 0 & 0 & T_{33} & T_{34} \\ 0 & 0 & 0 & T_{44} \end{bmatrix} = \begin{bmatrix} 1 & 0 & 0 & 0 \\ 0 & 1 & 0 & 0 \\ 0 & 0 & 1 & 0 \\ 0 & 0 & 0 & 1 \end{bmatrix} \quad (1\text{-}24)$$

Since the elements of the matrix $[T]$ are already determined using the elements of the known matrix $[A]$, then the $[S]$ matrix can be determined by performing the multiplication operation. Again, a few steps of the multiplication are performed for illustration.

$$S_{11} \cdot T_{11} = 1; \quad S_{11} = \frac{1}{T_{11}}$$

$$S_{11}T_{12} + S_{12}T_{22} = 0; \quad S_{12} = \frac{-S_{11}T_{12}}{T_{22}} = \frac{-T_{12}}{T_{11}T_{22}}$$

$$S_{11}T_{13} + S_{12}T_{23} + S_{12}T_{33} = 0; \quad S_{13} = \frac{-(S_{11}T_{13} + S_{12}T_{23})}{T_{33}}$$

$$S_{11}T_{14} + S_{12}T_{24} + S_{13}T_{34} + S_{14}T_{44} = 0; \quad S_{14} = \frac{-(S_{11}T_{14} + S_{12}T_{24} + S_{13}T_{34})}{T_{44}}$$

$$S_{22} \cdot T_{22} = 1; \quad S_{22} = \frac{1}{T_{22}}$$

This expansion can be generalized by a mathematical expression where i represents the row and j the column locations of the elements.

$$S_{ij} = -\sum_{k=i}^{j-1} \frac{S_{ik}T_{kj}}{T_{jj}} \quad \text{for} \quad j > i \quad (1\text{-}25)$$

and

$$S_{ii} = \frac{1}{T_{ii}} \quad \text{for} \quad j = i \quad (1\text{-}26)$$

The determination of the matrix $[S]$ which is the inverse of the upper triangular matrix $[T]$ has the intermediate duty of being used for the computation of the inverse of the original matrix $[A]$. This is done by the following logical process.

$$[A] = [L][T]$$

Postmultiply both sides of the matrix equation by $[S]$:

$$[A][S] = [L][T][S] = [L]$$

Premultiply both sides of the matrix equation by $[A]^{-1}$:

$$[A]^{-1}[A][S] = [A]^{-1}[L]$$
$$[S] = [A]^{-1}[L] \tag{1-27}$$

Since all the elements of the matrices $[S]$ and $[L]$ are known, the elements of $[A]^{-1}$ can be readily obtained by performing the matrix multiplication. If one lets $[A]^{-1}$ be $[X]$, then, the following matrix expression is written:

$$\begin{bmatrix} S_{11} & S_{12} & S_{13} & S_{14} \\ 0 & S_{22} & S_{23} & S_{24} \\ 0 & 0 & S_{33} & S_{34} \\ 0 & 0 & 0 & S_{44} \end{bmatrix} = \begin{bmatrix} X_{11} & X_{12} & X_{13} & X_{14} \\ X_{21} & X_{22} & X_{23} & X_{24} \\ X_{31} & X_{32} & X_{33} & X_{34} \\ X_{41} & X_{42} & X_{43} & X_{44} \end{bmatrix} \cdot \begin{bmatrix} 1 & 0 & 0 & 0 \\ L_{21} & 1 & 0 & 0 \\ L_{31} & L_{32} & 1 & 0 \\ L_{41} & L_{42} & L_{43} & 1 \end{bmatrix}$$
(1-28)

For simplicity, the multiplication is performed using a reverse order. A few steps of this operation are given below which yield the elements of matrix $[X] = [A]^{-1}$:

$S_{14} = X_{14}$
$S_{13} = X_{13} + L_{43} \cdot X_{14}; \quad X_{13} = S_{13} - L_{43} S_{14}$
$S_{12} = X_{12} + L_{32} \cdot X_{13} + L_{42} X_{14}; \quad X_{12} = S_{12} - L_{32} X_{13} - L_{42} X_{14}$
$S_{11} = X_{11} + X_{12} \cdot L_{21} + X_{13} \cdot L_{31} + X_{14} \cdot L_{41}$
$X_{11} = S_{11} - X_{12} \cdot L_{21} - X_{13} \cdot L_{31} - X_{14} \cdot L_{41}$
$S_{44} = X_{44}$

Cholesky's method offers additional advantages for matrices which are symmetric. A symmetric matrix $[A]$ can be written as the product of two triangular matrices, one of them being the transpose of the other. Now $[A] = [T]^T[T]$, which can be expanded into

MATRIX INVERSION

$$\begin{bmatrix} A_{11} & A_{12} & A_{13} & A_{14} \\ A_{21} & A_{22} & A_{23} & A_{24} \\ A_{31} & A_{32} & A_{33} & A_{34} \\ A_{41} & A_{42} & A_{43} & A_{44} \end{bmatrix} = \begin{bmatrix} T_{11} & 0 & 0 & 0 \\ T_{12} & T_{22} & 0 & 0 \\ T_{13} & T_{23} & T_{33} & 0 \\ T_{14} & T_{24} & T_{34} & T_{44} \end{bmatrix} \cdot \begin{bmatrix} T_{11} & T_{12} & T_{13} & T_{14} \\ 0 & T_{22} & T_{23} & T_{24} \\ 0 & 0 & T_{33} & T_{34} \\ 0 & 0 & 0 & T_{44} \end{bmatrix}$$

(1-29)

The elements of the matrix $[T]$ are obtained by the multiplication of the matrices of Equation (1-29):

$$A_{11} = T_{11}^2; \quad T_{11} = A_{11}^{1/2}$$

$$A_{12} = T_{11} \cdot T_{12}; \quad T_{12} = A_{12} \cdot \frac{1}{T_{11}} = A_{12}/A_{11}^{1/2}$$

$$A_{13} = T_{11} \cdot T_{13}; \quad T_{13} = A_{13} \cdot \frac{1}{T_{11}} = A_{13}/A_{11}^{1/2}$$

$$A_{14} = T_{11} \cdot T_{14}; \quad T_{14} = A_{14} \cdot \frac{1}{T_{11}} = A_{14}/A_{11}^{1/2}$$

$$A_{22} = T_{12}^2 + T_{22}^2; \quad T_{22} = (A_{22} - T_{12}^2)^{1/2} = (A_{22} - A_{12}^2/A_{11})^{1/2}$$

The above operations can be generalized by the mathematical expression

$$T_{ij} = \frac{A_{ij} - \sum_{k=1}^{i-1} T_{ki} T_{kj}}{T_{ii}} \quad \text{for } j \geq i$$

$$T_{11} = \sqrt{A_{11}}; \quad T_{1j} = \frac{A_{1j}}{T_{11}} \quad \text{for } i = 1 \quad (1\text{-}30)$$

When all the elements of $[T]$ are computed, then the inverse of the symmetric matrix $[A]$ is obtained by the operations defined in Equation (1-31), since the inverses of $[T]$ and $[T]^T$ are readily computed.

$$[A]^{-1} = [T]^{-1}[T^T]^{-1} \quad (1\text{-}31)$$

Example 1-9

Find the inverse of the given symmetric matrix $[A]$ by Cholesky's scheme.

$$[A] = \begin{bmatrix} 2 & 1 & 1 \\ 1 & 4 & 3 \\ 1 & 3 & 5 \end{bmatrix}$$

Solution

1 Determination of the elements of matrix $[T]$:

$$T_{11} = \sqrt{2} = 1.414$$

$$T_{12} = \frac{1}{\sqrt{2}} = 0.707$$

$$T_{13} = \frac{1}{\sqrt{2}} = 0.707$$

$$T_{22} = \sqrt{4 - (0.707)^2} = 1.87$$

$$T_{23} = \frac{3}{1.87} - \left(\frac{0.707}{1.87} \times 0.707\right) = \frac{2.5}{1.87} = 1.34$$

$$T_{33} = \sqrt{5 - (0.707 \times 0.707) + (1.34 \times 1.34)} = 1.64$$

Then the matrix $[T]$ is

$$[T] = \begin{bmatrix} 1.414 & 0.707 & 0.707 \\ 0 & 1.87 & 1.34 \\ 0 & 0 & 1.64 \end{bmatrix}$$

2 Determination of the inverse of the matrix $[T]$, namely $[T]^{-1} = [S]$

$$S_{11} = \frac{1}{1.414} = 0.707$$

$$S_{22} = \frac{1}{1.87} = 0.535$$

$$S_{33} = \frac{1}{1.64} = 0.61$$

$$S_{12} = -\frac{0.707}{1.87} \times 0.707 = -0.268$$

$$S_{13} = -\frac{0.707}{1.64} \times 0.707 + \frac{0.268}{1.64} \times 1.34 = -0.086$$

$$S_{23} = -\frac{0.535}{1.64} \times 1.34 = -0.327$$

Then the matrix $[S] = [T]^{-1}$ is

$$[S] = [T]^{-1} = \begin{bmatrix} 0.707 & -0.268 & -0.086 \\ 0 & 0.535 & -0.327 \\ 0 & 0 & 0.61 \end{bmatrix}$$

3 Determination of $[T]^T$ and $[T^T]^{-1}$

$$[T]^T = \begin{bmatrix} 1.414 & 0 & 0 \\ 0.707 & 1.87 & 0 \\ 0.707 & 1.34 & 1.64 \end{bmatrix}$$

and

$$[T^T]^{-1} = \begin{bmatrix} 0.707 & 0 & 0 \\ -0.268 & 0.535 & 0 \\ -0.086 & -0.327 & 0.61 \end{bmatrix}$$

4 Determination of the inverse of $[A]$ by Equation (1-31). (The numerical computations are performed by slide rule, so that the accuracy is not perfect.)

$$[A]^{-1} = [T]^{-1}[T^T]^{-1}$$

$$[A]^{-1} = \begin{bmatrix} 0.707 & -0.268 & -0.086 \\ 0 & 0.535 & -0.327 \\ 0 & 0 & 0.61 \end{bmatrix} \begin{bmatrix} 0.707 & 0 & 0 \\ -0.268 & 0.535 & 0 \\ -0.086 & -0.327 & 0.61 \end{bmatrix}$$

$$[A]^{-1} = \begin{bmatrix} 0.58 & -0.11 & -0.05 \\ -0.11 & 0.48 & -0.27 \\ -0.05 & -0.27 & 0.37 \end{bmatrix}$$

Example 1-10

Find the inverse of the given matrix $[A]$ whose elements are polynomials. If it has no inverse, explain why.

$$(A) = \begin{bmatrix} x+1 & x & 1 \\ -x^2+5 & -x^2-x+5 & x \\ -x & 0 & -x \end{bmatrix}$$

$$|A| = -x(x^2 + x^2 + x - 5) - x(x+1)(-x^2 - x + 5) - x(-x^2 + 5)$$
$$= -2x^3 - x^2 + 5x - 2x^3 + x^2 - 5x = 0$$

Since the determinant of $[A]$ is zero, this matrix has no inverse.
A matrix whose determinant is zero is known as a *singular matrix*.

Example 1-11

There are applications of matrix operations in vector algebra. The scalar triple product in vector analysis is defined as $A \cdot (B \times C)$ where A, B, and C are vector quantities. This quantity defines the volume of a parallelepiped. Then one can state that if $A \cdot (B \times C)$ is positive, we have a positive volume, meaning that the three vector quantities form a positive set. If $A \cdot (B \times C)$ is zero, then we have no volume, so that the three vectors are obviously coplanar. Given three vectors

$$\bar{a} = (3, 3, -1); \quad \bar{b} = (1, 1, 5); \quad \bar{c} = (16, -11, -1)$$

make a study of their triple product and define their property.

$$\bar{a} \cdot (\bar{b} \times \bar{c}) = \begin{vmatrix} 3 & 3 & -1 \\ 1 & 1 & 5 \\ 16 & -11 & -1 \end{vmatrix}$$

$$= 3(-1 + 55) - 3(-1 - 80) - 1(-11 - 16)$$
$$= 162 + 243 + 27 = 432$$

Since $\bar{a} \cdot (\bar{b} \times \bar{c}) = 432 \neq 0$, the set of $\bar{a}, \bar{b}, \bar{c}$ form a positive set.

COMMENTS ON MATRIX OPERATIONS

The state of the art in the use of digital computers for the solutions of problems in structural analysis demands the improvement of the applications of known matrix operations. The solution of a system of linear equations related to linear elastic analysis comprises the major part of the problem-solving effort. Statistical information indicates that almost one half of the computer time involved in structural analysis is related to the solution of simultaneous equations. Hence the improvement of the equation-solving process will improve the total solution system.

The matrix quantities which will be encountered in the displacement structural analysis method are generally varied; however, the stiffness matrix itself is symmetric, positive-definite, and usually banded. All these qualities can be used to increase the efficiency of the solution process by decreasing the computer time as well as the required computer storage space. The diagonal elements and elements above or below the diagonal of a symmetric matrix need to be stored; a positive-definite matrix requires less computational steps and a well-banded matrix has the majority of the elements close to its diagonal. The bandwidth is an important criteria for the quick solution of a system, since the solution time varies approximately as the square of the bandwidth.

This chapter introduced several schemes of matrix inversions which can be used for the solution of simultaneous equations. They are

1. Adjoint method
2. Gauss-Jordan method
3. Partitioning method
4. Cholesky's method

The adjoint method has an academic value more than a practical one. It is not meant to be incorporated in sophisticated computer programs. The Gauss-Jordan method is a systematic elimination process involving a minimum number of computational steps and is a widely used algorithm in general computer programs. The partitioning method is desirable for operating on a large matrix by parts. On special occasions it is very efficient and desirable. Cholesky's method, which deals with triangular matrices, requires less computer storage space. Its definition includes square-root operations, which decreases the accuracy of this method. When computers with word lengths of 36 bits or less are used, double precision is needed to insure accurate results.

There are other schemes for solving simultaneous equations which are published in scientific periodicals [3]. It is appropriate to mention that there is a method [4] which combines the best qualities of Cholesky and Gauss-Jordan algorithms, introducing a valuable approach for equation solving.

PROBLEMS

1-1 Add the following three matrices.

$$\begin{bmatrix} a_{11} & a_{12} & a_{13} \\ a_{21} & a_{22} & a_{23} \\ a_{31} & a_{32} & a_{33} \end{bmatrix} + \begin{bmatrix} 1 & 5 & 3 \\ 9 & 11 & 15 \\ 0 & 7 & 8 \end{bmatrix} + \begin{bmatrix} 0 & 9 & 19 \\ 15 & 22 & 10 \\ 3 & 2 & 1 \end{bmatrix}$$

1-2 What are the conditions for adding, equating, and multiplying two matrices? Give numerical examples and explain.

1-3 Multiply the given matrices:

a.
$$\begin{bmatrix} a_{11} & a_{12} & a_{13} \\ a_{21} & a_{22} & a_{23} \end{bmatrix} \begin{bmatrix} b_{11} & b_{12} \\ b_{21} & b_{22} \\ b_{31} & b_{32} \end{bmatrix} =$$

b. $\begin{bmatrix} 1 & 15 \\ 9 & 21 \end{bmatrix} \begin{bmatrix} 0 & 13 \\ 15 & 1 \end{bmatrix} =$

c. $\begin{bmatrix} 1 & 2 & 3 \\ 9 & 0 & 7 \\ 11 & 15 & 4 \end{bmatrix} \begin{bmatrix} 11 & 0 & 1 \\ 2 & 3 & 4 \\ 5 & 9 & 7 \end{bmatrix} =$

1-4 Find the determinant of the given matrix.

$$[A] = \begin{bmatrix} 4 & 7 & 9 \\ 11 & 0 & 13 \\ 15 & 2 & 1 \end{bmatrix}$$

1-5 Invert the given matrix by the Gauss-Jordan method.

$$[A] = \begin{bmatrix} 11 & 2 & 3 \\ 5 & 9 & 7 \\ 21 & 0 & 14 \end{bmatrix}$$

1-6 Invert the given matrix by partitioning.

$$[A] = \begin{bmatrix} 4 & 11 & 0 \\ 2 & 3 & 1 \\ 1 & 7 & 2 \end{bmatrix}$$

1-7 Invert the given matrix by the adjoint method.

$$[B] = \begin{bmatrix} 12 & 1 & 7 \\ 9 & 6 & 5 \\ 1 & 0 & 1 \end{bmatrix}$$

1-8 Invert the given matrix by pivotal condensation.

$$[C] = \begin{bmatrix} 11 & 15 & 1 \\ 13 & 21 & 4 \\ 5 & 2 & 3 \end{bmatrix}$$

1-9 Invert the given matrix $[A]$ by the Gauss-Jordan method, by the partitioning method and by the adjoint method. Then compare the necessary work to perform them.

$$[A] = \begin{bmatrix} 1 & 5 & 2 \\ 3 & 6 & 9 \\ 2 & 0 & 1 \end{bmatrix}$$

1-10 Solve the given system of linear equations by matrix operations.

a. $2x + 5y = 11$
$13x + 9y = 25$

b. $x + 3y + 11z = -4$
$2x + y + 4z = 12$
$3x + 2y + z = 10$

1-11 Solve the systems of equations given in Problem 1-10 by Cholesky's scheme and compare the result.

1-12 Explain the importance of matrix operations in the study of structural analysis.

1-13 For the matrices $[A]$ and $[B]$ given below, compute $[AB]^T$ and $[B^T][A^T]$, independently of each other, and show that $[AB]^T = B^T A^T$.

$$A = \begin{bmatrix} 1 & 2 & 3 \\ 4 & 5 & 6 \\ 7 & 8 & 9 \end{bmatrix}; \quad B = \begin{bmatrix} 10 & 13 \\ 11 & 14 \\ 12 & 15 \end{bmatrix}$$

REFERENCES

1. E. T. BELL, *Men of Mathematics.* New York: Simon and Schuster, 1937.
2. P. D. CROUT, "A Short Method for Evaluating Determinants and Solving Systems of Linear Equations with Real or Complex Coefficients," *Transactions of the AIEE* **60**, 1235, (1941).
3. B. M. IRONS, "A Frontal Solution Program for Finite-Element Analysis," *International Journal for Numerical Methods in Engineering* **2**, 1 (January-March, 1970).
4. R. J. MELOSH, "Manipulation Errors in Finite Element Analysis," NASA Contractors Report No. 1385, Philco-Ford, 1968.

Matrix Methods of Structural Analysis

2

INTRODUCTION

The knowledge of matrix concepts and operations with the availability of high-speed digital computers has opened new horizons for structural analysis. Methods such as moment distribution, slope-deflection, three-moment equations, elastic center, and column analogy have lost some of their glamor in practical applications to matrix structural analysis methods. The underlying basic concepts of all the analysis methods are the laws of mechanics themselves. Linear structural systems need to obey the following conditions: equilibrium of forces, compatibility of deformations, and boundary conditions.

The recent rapid development and success of matrix methods of structural analysis can be attributed to a great need for more accurate and faster solutions to complex structural problems. Matrix notation is well suited to the formulation of any type of structural analysis problem. An engineer who is familiar with matrix concepts can effectively and efficiently express the interrelations of his structural system. Since matrix operations are readily performed by the electronic digital computer, the engineer can obtain the solutions of his analysis in a very short time compared to that required by longhand methods.

The classical methods of analysis were applicable to comparatively simple elastic system with one-dimensional members. The matrix structural analysis methods become more useful and more efficient when applied to large and complex structural analysis problems. Structural systems with two- and three-dimensional elements can also be formulated and analyzed by matrix structural methods without great difficulty. At a time when many structural analysis and design problems are becoming complex and large, engineers are fortunate to have these modern methods at their service.

Matrix structural analysis methods can be classified into two categories, depending on the nature of the unknowns of the problem. If the forces of the structural system are the unknowns of the problem, then the force method (flexibility matrix method) is the approach to be used. If the displacements of the system are the choice for unknowns of the problem, the displacement method (stiffness matrix method) is the one to be used. Both of these methods are related to each other, for both must satisfy the basic requirements of structural mechanics.

This duality in structural analysis methods is also present in the study of the theory of elasticity. In the case of the stress approach of analysis of a two-dimensional body without body forces, the three stress values σ_x, σ_y, and τ_{xy} must satisfy the equilibrium equations

$$\frac{\partial \sigma_x}{\partial x} + \frac{\partial \tau_{xy}}{\partial y} = 0$$

and (2-1)

$$\frac{\partial \tau_{xy}}{\partial x} + \frac{\partial \sigma_y}{\partial y} = 0$$

Since one more equation is required to obtain the solution, the compatibility condition will yield the additional equation as a function of strains ϵ_x, ϵ_y, and γ_{xy}.

$$\frac{\partial^2 \epsilon_x}{\partial y^2} + \frac{\partial^2 \epsilon_x}{\partial x^2} = \frac{\partial^2 \gamma_{xy}}{\partial x \partial y} \qquad (2\text{-}2)$$

To correlate these three equations and make them all functions of stresses, Hooke's law is applied.

$$\epsilon_x = \frac{1}{E}(\sigma_x - \nu \sigma_y)$$

$$\epsilon_y = \frac{1}{E}(\sigma_y - \nu \sigma_x)$$

$$\gamma_{xy} = \frac{\tau_{xy}}{G} \tag{2-3}$$

This particular logic in the theory of elasticity can be considered as similar to the flexibility matrix method of analysis.

In the case of the displacement approach of analysis, the two equilibrium equations (2-1) can be rewritten as functions of linear displacements u in the x direction and v in y direction, using the strain-displacement relations:

$$\epsilon_x = \frac{\partial u}{\partial x}; \quad \epsilon_y = \frac{\partial v}{\partial y}; \quad \gamma_{xy} = \frac{\partial u}{\partial y} + \frac{\partial v}{\partial x} \tag{2-4}$$

The resulting two equations in u and v will yield the displacement values.

$$\frac{\partial^2 u}{\partial x^2} + \frac{\partial^2 u}{\partial y^2} = \frac{1+\nu}{2}\left(\frac{\partial^2 u}{\partial y^2} - \frac{\partial^2 v}{\partial x \partial y}\right)$$

$$\frac{\partial^2 v}{\partial x^2} + \frac{\partial^2 v}{\partial y^2} = \frac{1+\nu}{2}\left(\frac{\partial^2 v}{\partial x^2} - \frac{\partial^2 u}{\partial x \partial y}\right) \tag{2-5}$$

The stresses can then be computed by the help of Equations (2-3) and (2-4). This particular approach in the theory of elasticity can be considered as similar to the stiffness matrix method of analysis.

The matrix methods of structural analysis have important value in the scope of this book, since they are the basis of the development of the finite-element method of structural analysis. Hence a solid knowledge of matrix methods will ensure the complete and easy understanding of the finite-element method. This chapter will introduce the necessary concepts of matrix structural analysis methods in a concise but complete manner.

Let us redefine a few basic definitions of structural theory. A structural system is statically determinate if all the unknowns can be obtained by the equilibrium conditions. In the case of static indeterminacy, additional conditions—namely, the compatibility conditions—are needed to generate the other necessary equations. A system is kinematically determinate if all the possible displacements occur at the actual load points and in the direction of these loads.

A. FLEXIBILITY MATRIX METHOD

This method is also known as the *force method* since the forces are selected to be the unknowns of the problem in the case of indeterminate structural analysis. The basic principles to be satisfied are the equilibrium and compatibility conditions of the structural system with the observance of

36 MATRIX METHODS OF STRUCTURAL ANALYSIS

Hooke's law. In the case of determinate structural systems, the equilibrium equations are sufficient to obtain the solution of the problem. This approach will generate the necessary system of linear equations written at nodes and as functions of redundant member forces. The logic in the development of the method for a determinate structure is as follows.

Statically Determinate System

a. $\{P\} = [A]\{Q\}$ (2-6)

This expression relates the member-end forces $\{P\}$ to external loads $\{Q\}$ by a load-force matrix $[A]$. The equilibrium conditions of all forces at the nodes are the source of the development of matrix equation (2-6). The elements A_{ij} of the matrix, $[A]$, are force influence coefficients defined as the force at a point i produced by a unit load at a point j.

b. $\{U\} = [B]\{P\}$ (2-7)

This matrix equation relates the member deformations $\{U\}$ to member end forces $\{P\}$ by a force-deformation matrix $[B]$. The elements, B_{ij} of the matrix $[B]$ are deformation influence coefficients defined as the deformation at a point i produced by a unit force at a point j.

c. $\{D\} = [C]\{U\}$ (2-8)

This matrix equation will yield node displacements $\{D\}$ as a function of member-end deformations $\{U\}$, which are operated on by a deformation-displacement matrix $[C]$. The elements C_{ij} of the matrix $[C]$ are displacement coefficients defined as the displacement at a point i produced by a unit deformation at a point j when all other deformations are restrained.

The deformation-displacement matrix $[C]$ is the same as the transpose of the load-force matrix $[A]$.

$$[C] = [A]^T \quad (2\text{-}9)$$

This relation can be shown to be true by the following logical process. A structural system in equilibrium will generate the matrix equilibrium equation (2-6):

$$\{P\} = [A]\{Q\}$$

The applied loads $\{Q\}$ will cause elastic deformations of the members of the structure $[U]$, which in their turn generate node displacements $[D]$. Let us induce virtual displacements $\{\Delta D\}$ at the load points of the structural system. They will create virtual deformations $\{\Delta U\}$ in the members. The principle of conservation of energy will require that the external work done by the actual loads $\{Q\}$ moving through the virtual displacements

$\{\Delta D\}$ be equal to the internal work done by the member forces $\{P\}$ as the members suffer virtual deformations $\{\Delta U\}$. In mathematical notation, this relation is written as

$$\{P\}^T\{\Delta U\} = \{Q\}^T\{\Delta D\} \tag{2-10}$$

Using the previously defined relations, $\{\Delta D\} = [C] \cdot \{\Delta U\}$ and $[P]^T = \{[A]\{Q\}\}^T = \{Q\}^T[A]^T$, expression (2-10) is rewritten,

$$\{P\}^T\{\Delta U\} = \{Q\}^T[C]\{\Delta U\} \tag{2-11}$$

$$\{Q\}^T[A]^T\{\Delta U\} = \{Q\}^T[C]\{\Delta U\} \tag{2-12}$$

Therefore, one can conclude that $[A]^T$ is equal to $[C]$.

In the case of determinate structures, if the only desired values are the member forces, then Equation (2-6) will be sufficient. However, the node displacements can also be computed by the use of Equations (2-7) and (2-8).

The transformation matrices $[A]$, $[B]$, and $[C]$ are interrelated to develop a new matrix relation to correlate the nodal displacements with nodal forces. This is done by proper matrix quantity substitutions, as illustrated by the following operations.

$$\begin{aligned} \{D\} &= [C]\{U\} = [A]^T\{U\} \\ \{D\} &= [A]^T[B]\{P\} \end{aligned} \tag{2-13}$$

$$\{D\} = [A]^T[B][A]\{Q\} \tag{2-14}$$

$$\{D\} = [F]\{Q\} \tag{2-15}$$

The three matrix quantities which relate the nodal forces to nodal displacements is known as the *flexibility matrix* $[F]$.

$$[F] = [A]^T[B][A] \tag{2-16}$$

The element f_{ij} of the flexibility matrix $[F]$ is defined as the displacement at a point i, when a unit load is applied at a point j of the structure. Observing the characteristics of the flexibility matrix, one can reach the following conclusions:

1. Due to the similar dimensions of the matrix quantities, $\{D\}$ and $\{P\}$, the flexibility matrix is a square matrix.
2. Due to the application of Maxwell-Betti reciprocity theorem of elastic systems, the flexibility matrix is a symmetric system, $\{f_{ij}\} = \{f_{ji}\}$.
3. Since the diagonal element $\{f_{ii}\}$ of the flexibility matrix, $[F]$, which is the displacement at i, in the direction of the load applied at the same point always has a positive value, then the flexibility

matrix is known as a positive definite matrix. All the diagonal elements of such a matrix are positive.

Statically Indeterminate System

The use of the flexibility matrix method for the analysis of determinate structures is more academic in nature; however, this method becomes desirable and efficient when applied to the solution of indeterminate structures. The system of linear equations generated by the equilibrium conditions is not sufficient to resolve the analysis problem. Hence additional conditions—namely, the compatibility conditions—are needed to develop the additional equations. The following approach is used to formulate the indeterminate analysis problem by the flexibility matrix method.

a Remove redundant forces from the system until the structure becomes determinate. The selection of the redundant forces should be such that the stability of the system is not disturbed. The created determinate and stable structure is known as the primary structure. A load-force relation is written for the primary structure with the application of the actual loads.

$$\{P_Q\} = [A]\{Q\} \qquad (2\text{-}17)$$

where $\{P_Q\}$ is the member force matrix due to the externally applied loads.

b The redundant forces which are the unknowns of the analysis problem, are applied to the primary structure as loads. Then another load-force relation is obtained:

$$[P_x] = [G]\{X\} \qquad (2\text{-}18)$$

where $[P_x]$ is the member-force matrix due to redundant forces; $[G]$ is the load-force transformation matrix similar to matrix $[A]$ in nature, and $\{X\}$ is the redundant forces matrix. A logical matrix relations approach is developed with the same routine used for the determinate case.

$$[P] = \{P_Q\} + \{P_x\} \qquad (2\text{-}19)$$

$$\{P\} = [A]\{Q\} + [G]\{X\} \qquad (2\text{-}20)$$

$$\{U\} = [B]\{P\} \qquad (2\text{-}21)$$

$$\{U\} = [B][A:G]\begin{Bmatrix}Q\\ \cdots \\ X\end{Bmatrix} \qquad (2\text{-}22)$$

$$\{D\} = [A:G]^T\{U\} \qquad (2\text{-}23)$$

A. FLEXIBILITY MATRIX METHOD

$$\{D\} = [A:G]^T[B][A:G]\begin{Bmatrix} Q \\ \ddot{X} \end{Bmatrix} \quad (2\text{-}24)$$

The node displacements $\{D\}$ have contributions from the actual load $\{D_Q\}$ and from the redundant forces $\{D_x\}$. In matrix location, it can be written,

$$\{D\} = \begin{Bmatrix} D_Q \\ C_x \end{Bmatrix} = [A:G]^T[B][A:G]\begin{Bmatrix} Q \\ \ddot{X} \end{Bmatrix} \quad (2\text{-}25)$$

Equation (2-25) can be expanded into general matrix form by

$$\begin{Bmatrix} D_Q \\ \hline D_x \end{Bmatrix} = \begin{bmatrix} A^TBA & A^TBG \\ \hline G^TBA & G^TBG \end{bmatrix} \begin{Bmatrix} Q \\ \ddot{X} \end{Bmatrix} \quad (2\text{-}26)$$

Equation (2-26) can be rearranged to yield expressions for displacements due to actual loads and redundant forces:

$$\{D_Q\} = [A^TBA]\{Q\} + [A^TBG]\{X\} \quad (2\text{-}27)$$
$$\{D_x\} = [G^TBA]\{Q\} + [G^TBG]\{X\} \quad (2\text{-}28)$$

It was mentioned that the equilibrium conditions were not sufficient to generate enough number of equations to solve the analysis problem. The compatibility conditions are necessary to generate the lacking number of equations. The conditions to be used are the compatibility of the displacements due to the redundant forces $\{D_x\}$. These displacements $\{D_x\}$ can either be zero or equal to predetermined values. If one assumes that $\{D_x\} = 0$, then the boundary conditions will be satisfied by the equation

$$\{D_x\} = [G^TBA]\{Q\} + [G^TBG]\{X\} = 0 \quad (2\text{-}29)$$

Equation (2-29) can be rearranged to solve for the redundant forces which are the unknowns of the problem:

$$\{X\} = -[G^TBG]^{-1}[G^TBA]\{Q\} \quad (2\text{-}30)$$

The computation of the redundant forces $\{X\}$ will solve the major part of the problem. The values of $\{X\}$ are used to determine the complete member force matrix.

$$\{P\} = [A]\{Q\} + [G]\{X\} \quad (2\text{-}31)$$

If the nodal displacements of the statically indeterminate structure is needed, then the expression for $\{D_Q\}$ is used.

$$\{D_Q\} = [A^TBA]\{Q\} + [A^TBG]\{X\} \quad (2\text{-}32)$$

This ends the general formulation of structural analysis problems by the flexibility or force matrix method. Since all the expressions are written in matrix notation, they are readily computable by high-speed electronic digital computers. Everyone should be aware of the fact that digital computers are essential to perform large matrix operations. The input quantities for a basic computer program based on flexibility matrix concepts are $[A]$, $[G]$, $[B]$, and $[Q]$. The output quantities will be $[X]$, $[P]$, and $[D_Q]$. More elaborate programs are able to generate most of the input matrices from the geometry of the structural system and the elastic properties of the materials used in the system.

B. STIFFNESS MATRIX METHOD

This is also known as the *displacement method*, since the unknowns of the analysis problem are selected to be the displacements of the nodes. The compatibility conditions between node displacements and member deformations are initially established. The unknown node displacements are computed by the solution of a system of linear equations obtained through the application of the equilibrium conditions between the internal and external forces at the nodes. Since one-to-one correspondence exists between joint displacements and equilibrium conditions, the stiffness method can equally be applied to determinate and indeterminate structural analysis problems. As it is mentioned, the logic of this particular method is first to determine the unknown node displacements of the structural system and then compute the necessary member forces to complete the analysis. There are many textbooks which treat the stiffness matrix structural analysis method to great depth and detail [1]. Our intention is to give the necessary concise background of this method as transition to the finite element method.

Since the stiffness matrix method deals with the displacements as the primary unknowns, the concepts of kinematic determinacy or indeterminacy are relevant and will be used in the development of the method.

The member-end forces $\{P\}$ can be expressed as the summation of two values: $\{P_0\}$, member-end forces due to actual external loads when all displacements at the nodes are held at zero, and $\{P_1\}$, member-end forces due to the joint displacements.

$$\{P\} = \{P_0\} + \{P_1\} \tag{2-33}$$

The discussion of the flexibility method introduced relations between the external loads, member forces, member deformations, and node displacements and these relations were defined by Equations (2-6), (2-7), and (2-8).

Now we will develop similar relations according to the logic of the stiffness matrix method.

a. $\{P_1\} = [K]\{U\}$ (2-34)

This matrix equation relates the member-end forces $\{P_1\}$, to member deformations $\{U\}$ by a deformation-force matrix $[K]$. The matrix $[K]$ is a function of the elastic properties and the dimensional configurations of the structural system, where the elements K_{ij} of the matrix $[K]$, are member-end forces due to a unit applied deformation, all other deformations remaining zero.

b. $\{U\} = \{\psi\}\{D\}$ (2-35)

The member deformations $\{U\}$ are related to node displacements $\{D\}$ by a displacement-deformation matrix $\{\psi\}$, which is a function of the geometry of the structure.

c. $\{Q\} = \{\psi\}^T\{P_1\}$ (2-36)

The matrix equation (2-36) illustrates the relation between the load matrix $\{Q\}$, and member forces matrix $\{P_1\}$, by a force-load matrix, $\{\psi\}^T$ which is the transpose of the displacement-deformation matrix $\{\psi\}$.

It is beneficial to continue the discussion by first considering kinematically determinate structural systems.

Kinematically Determinate System

The three basic matrix equations which are relevant to the analysis are (2-34), (2-35), and (2-36). They are rewritten to illustrate the necessary substitution to generate the Equation (2-37).

$$\begin{aligned} \{P_1\} &= [K]\{U\} \\ \{U\} &= \{\psi\}\{D\} \\ \{Q\} &= \{\psi\}^T\{P_1\} \\ \{Q\} &= \{\psi\}^T[K]\{\psi\}\{D\} \end{aligned} \quad (2\text{-}37)$$

The matrix quantity $[\{\psi\}^T[K]\{\psi\}]$ is known as the stiffness matrix of a kinematically determinate structure. Let us call $[S]$ the structural stiffness matrix.

$$[S] = \{\psi\}^T[K]\{\psi\} \quad (2\text{-}38)$$

Then Equation (2-37) is rewritten as

$$\{Q\} = [S]\{D\} \quad (2\text{-}39)$$

which is manipulated to obtain the unknown node displacements of the problem.

$$\{D\} = [S]^{-1}\{Q\} \quad (2\text{-}40)$$

The computed values of the displacements are used in Equation (2-35) for the determination of member deformations $\{U\}$, and in Equation (2-34) for the determination of the member forces, $\{P_1\}$.

Kinematically Indeterminate System

A kinematically indeterminate system has two types of nodal displacements. First, displacements $\{D\}$ which occur in the direction of applied loads; second, displacements $\{\Delta\}$ which occur at points or/and directions where no loads are applied. The contributions of $\{D\}$ displacements to member deformations are given by a matrix $\{U_d\}$ and the contributions of $\{\Delta\}$ displacements to member deformations are given by a matrix $\{U_\Delta\}$. The total member deformations will be the sum of $\{U_d\}$ and $\{U_\Delta\}$.

$$\{U\} = \{U_d\} + \{U_\Delta\} \tag{2-41}$$

Using Equation (2-35), the above expression is rewritten as

$$\{U\} = \{\psi\}\{D\} + \{R\}\{\Delta\} \tag{2-42}$$

where the displacement-deformation matrix $\{R\}$ is generated in a similar manner as previously defined matrix $\{\psi\}$. The general equation (2-34) is written to describe the kinematic indeterminacy condition:

$$\{P_1\} = \{P_d\} + \{P_\Delta\} \tag{2-43}$$

The matrix $\{P_d\}$ is the member forces induced by node displacements $\{D\}$ and the matrix $\{P_\Delta\}$ is the member forces induced by node displacements $\{\Delta\}$.

In the case of loads applied at the nodes, the member force matrix $\{P_0\}$ of Equation (2-33) will vanish. Then the development of the stiffness method will have the following pattern.

$$\{Q\} = \{\psi\}^T\{P_1\} = \{\psi\}^T[K]\{U\} \tag{2-44}$$

The substitution of $\{U\}$ from Equation (2-42) will yield

$$\{Q\} = \{\psi\}^T[K]\{\psi\}\{D\} + \{\psi\}^T[K]\{R\}\{\Delta\} \tag{2-45}$$

Assume a virtual load matrix $\{X\}$ applied in the directions of the displacements $\{\Delta\}$. Then an expression similar to Equation (2-45) can be written:

$$\{X\} = \{R\}^T[K]\{\psi\}\{D\} + \{R\}^T[K]\{R\}\{\Delta\} \tag{2-46}$$

B. STIFFNESS MATRIX METHOD

The expressions (2-45) and (2-46) are combined in a single matrix form as

$$\begin{Bmatrix} Q \\ X \end{Bmatrix} = \begin{Bmatrix} \psi^T \\ R^T \end{Bmatrix} [K]\{\psi:R\} \begin{Bmatrix} D \\ \Delta \end{Bmatrix} \qquad (2\text{-}47)$$

The definition of the stiffness matrix $\{S\}$ is given as

$$\{S\} = \begin{Bmatrix} \psi^T \\ R^T \end{Bmatrix} [K]\{\psi:R\} \qquad (2\text{-}48)$$

which simplifies the notation of Equation (2-47) as

$$\begin{Bmatrix} Q \\ X \end{Bmatrix} = [S] \begin{Bmatrix} D \\ \Delta \end{Bmatrix} \qquad (2\text{-}49)$$

Since the nodal displacements are required, Equation (2-49) is rearranged to allow the direction solution of the unknown displacements.

$$\begin{Bmatrix} D \\ \Delta \end{Bmatrix} = [S]^{-1} \begin{Bmatrix} Q \\ X \end{Bmatrix} \qquad (2\text{-}50)$$

The member deformations are readily computed by Equation (2-42) once the displacements values are determined from Equation (2-50). The member forces are calculated by Equation (2-43), which can be expanded into an expression as follows:

$$\{P_1\} = [K]\{\psi:R\}[S]^{-1} \begin{Bmatrix} Q \\ X \end{Bmatrix} \qquad (2\text{-}51)$$

During the final numerical computations, the virtual loads $\{X\}$ should be made zero, since they have no real values. The $\{P_1\}$ matrix is the final member forces matrix since $\{P_0\}$ is a null matrix due to the application of all loads at the nodes.

The case where loads are applied at points other than the nodes will require the consideration of $\{P_0\}$ member forces matrix. The procedure to obtain a solution to this problem will start by letting Equation (2-46) be equal to zero:

$$\{X\} = \{R\}^T[K]\{\psi\}\{D\} + \{R\}^T[K]\{R\}\{\Delta\} = 0$$

which is rearranged into the expression

$$\{R\}^T[K]\{\psi\}\{D\} = -\{R\}^T[K]\{R\}\{\Delta\} \qquad (2\text{-}52)$$

which in turn is solved for $\{\Delta\}$:

$$\{\Delta\} = -[\{R\}^T[K]\{R\}]^{-1}[\{R\}^T[K]\{\psi\}]\{D\} \qquad (2\text{-}53)$$

Equation (2-53) needs to be freed from the unknown node displacements $\{D\}$ in order to solve for $\{\Delta\}$ matrix. The final member forces $\{P\}$, in this case, will be

$$\{P\} = \{P_0\} + \{P_\Delta\} \qquad (2\text{-}54)$$

The matrix quantity $\{P_d\}$ is omitted since there are no applied loads at the nodes. Equation (2-54) can be rewritten in the form

$$\{P\} = \{P_0\} + [K]\{R\}\{\Delta\} \qquad (2\text{-}55)$$

The virtual load matrix $\{X\}$ is related to member-force matrix $\{P\}$ with the same logic displayed in Equation (2-44): $\{Q\}$ is the external node loads computed by the algebraic summation of the $\{P_0\}$ forces at each node.

$$\{X\} = \{R\}^T\{P\} = \{R\}^T\{Q\} + \{R\}^T[K]\{R\}\{\Delta\} \qquad (2\text{-}56)$$

Since we have established that $\{X\}$ is a null matrix, displacements $\{\Delta\}$ will be

$$\{\Delta\} = -[\{R\}^T[K]\{R\}]^{-1}\{R\}^T\{Q\} \qquad (2\text{-}57)$$

The comparison of Equation (2-53) with Equation (2-57) illustrates the elimination of $\{D\}$ which was prohibiting the direct computation of matrix $\{\Delta\}$.

The determination of the matrix $\{\Delta\}$ will permit the computations of the member forces matrix $\{P\}$. Substitute Equation (2-57) in Equation (2-55) to obtain Equation (2-58):

$$\{P\} = \{P_0\} - [K]\{R\}[\{R\}^T[K]\{R\}]^{-1}\{R\}^T\{Q\} \qquad (2\text{-}58)$$

This matrix expression can be rearranged for practical purposes into a new form:

$$\{P\} = \{[I] - [K]\{R\}[\{R\}^T[K]\{R\}]^{-1}\{R\}^T\}\{Q\} \qquad (2\text{-}59)$$

The member deformation matrix $\{U\}$ is computed by the use of Equation (2-42).

Let us pause for a moment to note the properties of the stiffness matrix $\{S\}$.

1. Because of one-to-one correspondence between the loads and displacements, the stiffness matrix is a square matrix.
2. Because of the Maxwell-Betti theorm of reciprocity, the stiffness matrix is a symmetric matrix.

3 Because of the consistency between the applied loads and resulting displacements (they have to have similar algebraic signs), the stiffness matrix is a positive definite matrix.

A. Comparative Study of Flexibility and Stiffness Matrix Methods:

Flexibility Matrix Method

1 The unknowns of the problem are the forces generated at the chosen releases to develop the primary structure.
2 For indeterminate structures, the compatibility conditions generate the additional necessary equations.
3 Flexibility method can solve the determinate structures using only the equilibrium conditions. For the indeterminate cases, the compatibility conditions are necessary.

Stiffness Matrix Method

1 The unknowns of the problem are the displacements.
2 Equilibrium conditions at nodes are used to generate the necessary equations.
3 The same theoretical steps are used for the solution of determinate and indeterminate analysis problems.

TABLE 2-1 RELATIONS BETWEEN MATRICES RELATED TO FLEXIBILITY AND STIFFNESS METHODS

Method	Force-Deformation Matrix	Load-Force Matrix	Deformation-Displacement Matrix	Load-Displacement Matrix
Stiffness	$P = K \cdot U$	$Q = \psi^T \cdot P$	$U = \psi \cdot D$	$Q = S \cdot D$
Flexibility	$U = B \cdot P$	$P = A \cdot Q$	$D = A^T \cdot U$	$D = F \cdot Q$
Relations between Matrices	$K = B^{-1}$ (for complete external matrices)	$\psi^T = A^{-1}$ (if both are square matrices)	$\psi^{-1} = A^T$ (if both are square matrices)	$S^{-1} = F$

Example 2-1

Perform the structural analysis of the three-span continuous beam shown in Fig. 2-1 by the stiffness matrix method. EI_c is constant.

Solution

1 The continuous beam is statically indeterminate to the second degree and kinematically indeterminate to the second degree. The

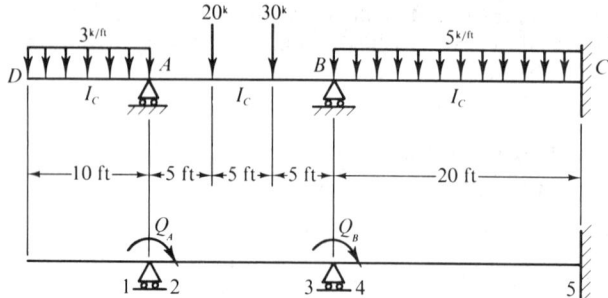

Figure 2-1

two unknown displacements are the rotational freedoms at joints A and B.

$$\{\Delta\} = \begin{Bmatrix} \theta_A \\ \theta_B \end{Bmatrix}$$

2   The member forces are given by Equation (2-54) in which $\{P_\Delta\}$ will be a function of only $\{\Delta\}$ displacements, since there is not a $\{D\}$ matrix for this problem.

$$\{P\} = \{P_0\} + \{P_\Delta\}$$

3   $\{P_0\}$ matrix is the member end forces due to external loads when all node displacements are zero. The second subscripts of the fixed-moment rotations describe the points where the moments are calculated.

$$P_{01} = -(30)(5) = -150 \text{ ft-kips}$$

$$P_{02} = \frac{(20)(5)(10)^2}{(15)^2} + \frac{(30)(10)(5)^2}{(15)^2} = 77.8 \text{ ft-kips}$$

$$P_{03} = -\frac{(20)(10)(5)^2}{(15)^2} - \frac{(30)(10)^2(5)}{(15)^2} = -88.8 \text{ ft-kips}$$

$$P_{04} = \frac{(5)(20)^2}{12} = 166.7 \text{ ft-kips}$$

$$P_{05} = -\frac{(5)(20)^2}{12} = -166.7 \text{ ft-kips}$$

4   The external node loads are computed by the algebraic summation of the $\{P_0\}$ forces at each node.

$$\{Q\} = \begin{Bmatrix} Q_A \\ Q_B \end{Bmatrix} = \begin{Bmatrix} (P_{01} + P_{02}) \\ (P_{03} + P_{04}) \end{Bmatrix} = \begin{Bmatrix} -72.2 \\ 77.9 \end{Bmatrix} \text{ ft-kips}$$

B. STIFFNESS MATRIX METHOD

5 Since we will be using Equation (2-55), the matrix quantities $[K]$ and $\{R\}$ are to be generated. First, let us develop the deformation force matrix $[K]$, using the beam deformation-force expression for a span AB. θ_A and θ_B are the rotations at nodes A and B. P_{AB} and P_{BA} are the member end forces due to occurring rotations.

$$P_{AB} = \frac{4EI}{L} \cdot \theta_A + \frac{2EI}{L} \theta_B$$

$$P_{BA} = \frac{2EI}{L} \theta_A + \frac{4EI}{L} \theta_B$$

According to Equation (2-34), one obtains the $[K]$ matrix for the problem. The units are feet for length and kips for force.

$$[K] = \begin{bmatrix} \frac{4EI_c}{15} & \frac{2EI_c}{15} & 0 & 0 \\ \frac{2EI_c}{15} & \frac{4EI_c}{15} & 0 & 0 \\ 0 & 0 & \frac{4EI_c}{20} & \frac{2EI_c}{20} \\ 0 & 0 & \frac{2EI_c}{20} & \frac{4EI_c}{20} \end{bmatrix} = EI_c \begin{bmatrix} 0.27 & 0.13 & 0 & 0 \\ 0.13 & 0.27 & 0 & 0 \\ 0 & 0 & 0.20 & 0.10 \\ 0 & 0 & 0.10 & 0.20 \end{bmatrix}$$

Secondly, let us develop the displacement-deformation matrix $\{R\}$ as needed for Equation (2-55). A general beam element will have the following possible forces and deformations, Fig. 2-2. The superscript denotes the member identification, the subscript 2 defines the rotation and moment at the left end of member and the subscript 3 defines the rotation and moment at the right end of the member. The axial deformations are neglected for all practical purposes. The matrix $\{R\}$ is computed by first letting $\theta_A = 1$ and $\theta_B = 0$ and then $\theta_A = 0$ and $\theta_B = 1$, which is allowed by the definition of displacement-deformation matrix

$$\{R\} = \begin{bmatrix} 1 & 0 \\ 0 & 1 \\ 0 & 1 \\ 0 & 0 \end{bmatrix}$$

6 Once the matrices $[K]$ and $\{R\}$ are generated, then Equation (2-57) is used to obtain the values of node displacements $\{\Delta\}$. Since there are several matrix operations, they will be performed in

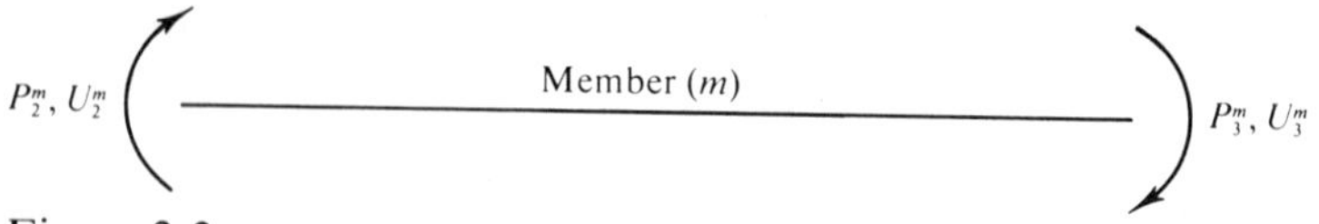

48 MATRIX METHODS OF STRUCTURAL ANALYSIS

P_2^m, U_2^m (——— Member (m) ———) P_3^m, U_3^m

Figure 2-2

parts and then collected for the simplicity of the solution procedure. The reader should correlate these operations to Equation (2-57).

Computation of $[R^TKR]^{-1}$

(a)

$$\{R^TK\} = EI_c \begin{Bmatrix} 1 & 0 & 0 & 0 \\ 0 & 1 & 1 & 0 \end{Bmatrix} \begin{bmatrix} 0.27 & 0.13 & 0 & 0 \\ 0.13 & 0.27 & 0 & 0 \\ 0 & 0 & 0.20 & 0.10 \\ 0 & 0 & 0.10 & 0.20 \end{bmatrix}$$

$$\{R^TK\} = EI_c \begin{Bmatrix} 0.27 & 0.13 & 0 & 0 \\ 0.13 & 0.27 & 0.2 & 0.1 \end{Bmatrix}$$

(b)

$$[R^TKR] = EI_c \begin{Bmatrix} 0.27 & 0.13 & 0 & 0 \\ 0.13 & 0.27 & 0.2 & 0.1 \end{Bmatrix} \begin{Bmatrix} 1 & 0 \\ 0 & 1 \\ 0 & 1 \\ 0 & 0 \end{Bmatrix} = EI_c \begin{Bmatrix} 0.27 & 0.13 \\ 0.13 & 0.47 \end{Bmatrix}$$

(c)

$$[R^TKR]^{-1} = \frac{1}{0.1101 EI_c} \cdot \begin{bmatrix} 0.47 & -0.13 \\ -0.13 & 0.27 \end{bmatrix}$$

Computation of $\{\Delta\} = -[R^TKR]^{-1}\{R\}^T\{Q\}$

By Equations (2-36) and (2-57), the $\{\Delta\}$ matrix is computed:

$$\{\Delta\} = \frac{-1}{0.1101 EI_c} \begin{bmatrix} 0.47 & -0.13 \\ -0.13 & 0.27 \end{bmatrix} \begin{Bmatrix} -72.2 \\ 77.9 \end{Bmatrix} = \frac{-1}{0.1101 EI_c} \begin{Bmatrix} -43.9 \\ 30.4 \end{Bmatrix}$$

7 The values of the member forces $\{P\}$ is obtained by Equation (2-55).

$$\{P\} = \{P_0\} + [K]\{R\}\{\Delta\}$$

$$\{P\} = \{P_0\} + EI_c \begin{bmatrix} 0.27 & 0.13 & 0 & 0 \\ 0.13 & 0.27 & 0 & 0 \\ 0 & 0 & 0.20 & 0.10 \\ 0 & 0 & 0.10 & 0.20 \end{bmatrix} \cdot \frac{-1}{0.1101 EI_c}$$

$$\cdot \begin{Bmatrix} 1 & 0 \\ 0 & 1 \\ 0 & 1 \\ 0 & 0 \end{Bmatrix} \begin{Bmatrix} -43.9 \\ 30.4 \end{Bmatrix}$$

$$\{P\} = \{P_0\} - \frac{1}{0.1101} \begin{Bmatrix} -7.88 \\ 2.50 \\ 6.08 \\ 3.04 \end{Bmatrix} = \begin{Bmatrix} 71.5 \\ -22.6 \\ -55.2 \\ -27.6 \end{Bmatrix}$$

8 The final member end forces are obtained by the algebraic sum of $\{P_0\}$ and $\{P_\Delta\}$.

$$\{P\} = \begin{Bmatrix} P_1 \\ P_2 \\ P_3 \\ P_4 \end{Bmatrix} = \begin{Bmatrix} 77.8 \\ -88.8 \\ 166.7 \\ -166.7 \end{Bmatrix} + \begin{Bmatrix} 71.5 \\ -22.6 \\ -55.3 \\ -27.6 \end{Bmatrix} = \begin{Bmatrix} 149.3 \\ -111.4 \\ 111.4 \\ -194.3 \end{Bmatrix} \text{ ft-kips}$$

Example 2-2

Perform the analysis of the structure shown in Fig. 2-3 by the stiffness matrix method. Generate the necessary matrices and then use the computer program at the end of Chapter 2 to solve the problem. E is constant.

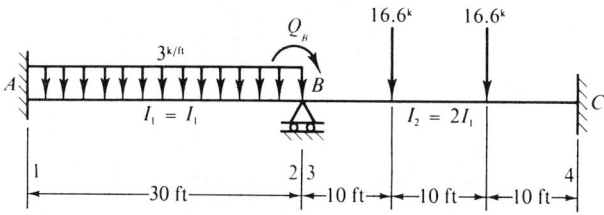

Figure 2-3

Solution:

1. The continuous beam is statically indeterminate to third degree and kinematically indeterminate to first degree. The displacement matrix $\{\Delta\}$ has only one element:

$$\{\Delta\} = \{\theta_B\}$$

2. Computation of the member-end forces due to the external loads when all rotations of the nodes are zero, $\{P_0\}$. The values of $\{P_0\}$ are computed by the general fixed-end moment expressions of structural analysis.[2]

$$\{P_{01}\} = \frac{(3)(30)^2}{12} = 225 \text{ ft-kips}$$

$$\{P_{02}\} = -225 \text{ ft-kips}$$

$$\{P_{03}\} = \frac{(16.6)(10)(20)^2}{(30)^2} + \frac{(16.6)(20)(10)^2}{(30)^2} = 110.7 \text{ ft-kips}$$

$$\{P_{04}\} = \frac{(16.6)(20)(10)^2}{(30)^2} + \frac{(16.6)(10)(20)^2}{(30)^2} = -110.7 \text{ ft-kips}$$

These values are summarized in matrix notation.

$$\{P_0\} = \begin{Bmatrix} P_{01} \\ P_{02} \\ P_{03} \\ P_{04} \end{Bmatrix} = \begin{Bmatrix} 225 \\ -225 \\ 110.7 \\ -110.7 \end{Bmatrix} \text{ ft-kips}$$

3. To compute the displacements and the member forces, Equations (2-55) and (2-57) are used. Since they include the matrix quantities $[K]$ and $\{R\}$, it is necessary to generate them. The matrix $[K]$ is generated with the same logic of Example 2-1:

$$[K] = EI_1 \begin{bmatrix} 0.1333 & 0.0667 & 0 & 0 \\ 0.0667 & 0.1333 & 0 & 0 \\ 0 & 0 & 0.2666 & 0.1333 \\ 0 & 0 & 0.1333 & 0.2666 \end{bmatrix}$$

The generation of the displacement-deformation matrix $\{R\}$ is readily performed by letting (θ_B) assume a value of unity while others are zero. Using the definition of the matrix $\{R\}$, one obtains

B. STIFFNESS MATRIX METHOD   51

$$\{R\} = \begin{Bmatrix} 0 \\ 1 \\ 1 \\ 0 \end{Bmatrix}$$

4   The matrix operations involved in the solutions of Equations (2-55) and (2-57) are performed by the aid of the computer program listed at the end of Chapter 2. The input quantities are given with the proper formats and the input matrices are read in by rows. At this time the reader should get acquainted with the mentioned program. The input matrices are defined at the end of the chapter.

Input Matrices

(a)   KIN = 1, NEF = 3, NLC = 1   FORMAT (3I5)
(b) R = GAMMA (I, J) FORMAT (8F10.4)
(c) K = $C(I,J)$ FORMAT (8F10.4)
(d) P_0 = $P_0(I,J)$ FORMAT (8F10.4)

Output Matrices

(a) S = $S(I, J)$ FORMAT (8F10.4)
(b) Δ = $D(I, J)$ FORMAT (8F10.4)
(c) P_Δ = $P_\Delta(I,J)$ FORMAT (8F10.4)
(d) P = $P(I,J)$ FORMAT (8F10.4)

5   The final member-end forces matrix of the structure is

$$\{P\} = \begin{Bmatrix} 243.9761 \\ -187.0762 \\ 187.0761 \\ -73.2762 \end{Bmatrix} \text{ft-kips}$$

6   The computer input and output matrices are shown in Figs. 2-4 and 2-5.

Example 2-3

Perform the analysis of the rigid frame shown in Fig. 2-6 by stiffness matrix method. Generate the necessary matrix quantities and then use the computer program given at the end of Chapter 2 to obtain the numerical solutions of the analysis. Let $E = 30 \times 10^6$ psi.

	FIELD IDENTIFICATION							
	1-10	11-20	21-30	31-40	41-50	51-60	61-70	71-80
jj 1	1.							
KIN, NEF, NLC, NJOIN 2	2.	4.						
$[\gamma]$ 3	0.	1.	0.	1.	0.	1.	0.	0.
$[k]$ 4	0...2.7	0...1.3	0.	0.	0...1.3	0...2.7	0.	0.
5	0.	0.	0...2	0...1	0.	0.	0...1	0...2
$[P_0]$ 6	7.7..8	−8.8..8	1.6.6..7	−1.6.6..7				
Control card 7	−1.							
8								
9								
10								
11								
12								
13								
14								
15								
16								
17								
18								
19								
20								

Figure 2-4

B. STIFFNESS MATRIX METHOD

| KIN= | 1 | NEF= | 4 | NLC= | 1 | NJOIN= | 0 |

DEFORMATION-DISPLACEMENT MATRIX--(GAMMA)

```
0.0000
1.0000
1.0000
0.0000
```

DEFORMATION-FORCE MATRIX--(K)

```
0.1333   0.0667   0.0000   0.0000
0.0667   0.1333   0.0000   0.0000
0.0000   0.0000   0.2667   0.1333
0.0000   0.0000   0.1333   0.2667
```

FIXED-END FORCES MATRIX

```
 225.0000
-225.0000
 111.2000
-111.2000
```

THE EXTERNAL STIFFNESS MATRIX--(S)

```
0.4000
```

THE REDUNDANT DISPLACEMENT MATRIX

```
284.4998
```

MEMBER-END FORCES DUE TO MEMBER-END DEFORMATIONS

```
18.9761
37.9238
75.8761
37.9238
```

TOTAL MEMBER END FORCES

```
 243.9761
-187.0762
 187.0761
 -73.2762
```

Figure 2-5

54 MATRIX METHODS OF STRUCTURAL ANALYSIS

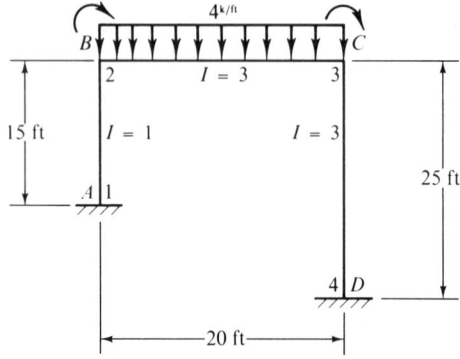

Figure 2-6

Solution

1. The rigid frame shown in Fig. 2-6 is statically indeterminate to third degree and also kinematically indeterminate to third degree. These are two degrees of freedom for rotations and one degree of freedom for translation. The displacement matrix $\{\Delta\}$ is written

$$\{\Delta\} = \begin{Bmatrix} \theta_B \\ \theta_C \\ \delta_{B,C} \end{Bmatrix}$$

2. The $\{P_0\}$ matrix is the member-end forces due to the external loads when all rotations and translations of the nodes are zero. The values of $\{P_0\}$ are computed by the general fixed-end moment expressions of structural analysis. Since there is no load on the vertical members of the frame, the only member with fixed-end moments will be the beam or top member. A typical frame member will have the following possible forces and deformations. Fig. 2-7. The superscripts denote the member identification and

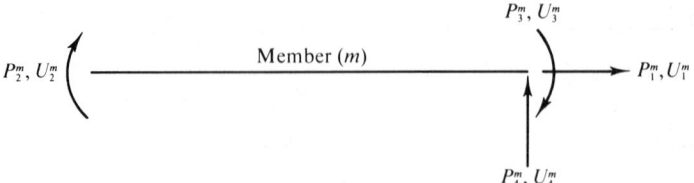

Figure 2-7

the subscripts denote the type of deformations and forces. According to this configuration of forces, $\{P_0\}$ is written:

B. STIFFNESS MATRIX METHOD

$$\{P_0\} = \begin{Bmatrix} P_{02}^1 \\ P_{03}^1 \\ P_{04}^1 \\ P_{02}^2 \\ P_{03}^2 \\ P_{02}^3 \\ P_{03}^3 \\ P_{04}^3 \end{Bmatrix} = \begin{Bmatrix} 0 \\ 0 \\ 0 \\ 133.33 \\ -133.33 \\ 0 \\ 0 \\ 0 \end{Bmatrix}$$

The axial force and deformation (P_1^m and U_1^m) are neglected due to the nature of the problem.

3 To compute the displacements $\{\Delta\}$ and the member forces, $\{P\}$, Equations (2-55) and (2-57) are used. Since they include the matrix quantities $[K]$ and $\{R\}$, it is necessary to generate them. The $[K]_m$ submatrix for one member is given by the general expression[3]

$$[K]_m = \begin{bmatrix} \dfrac{4EI}{L} & \dfrac{2EI}{L} & \dfrac{-6EI}{L^2} \\ \dfrac{2EI}{L} & \dfrac{4EI}{L} & \dfrac{-6EI}{L^2} \\ \dfrac{-6EI}{L^2} & \dfrac{-6EI}{L^2} & \dfrac{12EI}{L^3} \end{bmatrix} \qquad (2\text{-}60)$$

which relates the member-end forces of a member m to the resulting member-end deformations:

$$\begin{Bmatrix} P_2^m \\ P_3^m \\ P_4^m \end{Bmatrix} = \begin{bmatrix} \dfrac{4EI}{L} & \dfrac{2EI}{L} & \dfrac{-6EI}{L^2} \\ \dfrac{2EI}{L} & \dfrac{4EI}{L} & \dfrac{-6EI}{L^2} \\ \dfrac{-6EI}{L^2} & \dfrac{-6EI}{L^2} & \dfrac{12EI}{L^3} \end{bmatrix} \begin{Bmatrix} U_2^m \\ U_3^m \\ U_4^m \end{Bmatrix} \qquad (2\text{-}61)$$

In the case where the axial force (P_1^m) and axial deformation (U_1^m) are considered, then the complete deformation-force matrix

[K] for Fig. 2-5 can be written as

$$\begin{Bmatrix} P_1^m \\ P_2^m \\ P_3^m \\ P_4^m \end{Bmatrix} = \begin{bmatrix} \dfrac{AE}{L} & 0 & 0 & 0 \\ 0 & \dfrac{4EI}{L} & \dfrac{2EI}{L} & \dfrac{-6EI}{L^2} \\ 0 & \dfrac{2EI}{L} & \dfrac{4EI}{L} & \dfrac{-6EI}{L^2} \\ 0 & \dfrac{-6EI}{L^2} & \dfrac{-6EI}{L^2} & \dfrac{12EI}{L^3} \end{bmatrix} \begin{Bmatrix} U_1^m \\ U_2^m \\ U_3^m \\ U_4^m \end{Bmatrix} \qquad (2\text{-}62)$$

The units are feet for length, kips for force. The [K] matrix for the frame is obtained by using Equation (2-61),

$$[K] = \begin{bmatrix} 8000.0 & 4000.0 & -800.0 & 0 & 0 & 0 & 0 & 0 \\ 4000.0 & 8000.0 & -800.0 & 0 & 0 & 0 & 0 & 0 \\ -800.0 & -800.0 & 106.5 & 0 & 0 & 0 & 0 & 0 \\ 0 & 0 & 0 & 18000.0^* & 9000.0 & 0 & 0 & 0 \\ 0 & 0 & 0 & 9000.0 & 18000.0 & 0 & 0 & 0 \\ 0 & 0 & 0 & 0 & 0 & 14400.0 & 7200.0 & -864.0 \\ 0 & 0 & 0 & 0 & 0 & 7200.0 & 14400.0 & -865.0 \\ 0 & 0 & 0 & 0 & 0 & -864.0 & -864.0 & 69.1 \end{bmatrix}$$

The generation of the displacement-deformation matrix $\{R\}$ is performed by using the following three sets of displacement conditions and the definition of the matrix itself.

$$\begin{Bmatrix} U_2^1 \\ U_3^1 \\ U_4^1 \\ U_2^2 \\ U_3^2 \\ U_2^3 \\ U_3^3 \\ U_4^3 \end{Bmatrix} = \begin{bmatrix} 0 & 0 & 0 \\ 1 & 0 & 0 \\ 0 & 0 & 1 \\ 1 & 0 & 0 \\ 0 & 1 & 0 \\ 0 & 0 & 0 \\ 0 & 1 & 0 \\ 0 & 0 & 1 \end{bmatrix} \begin{Bmatrix} \theta_B \\ \theta_C \\ \delta_{B,C} \end{Bmatrix}$$

FIELD IDENTIFICATION

	1-10	11-20	21-30	31-40	41-50	51-60	61-70	71-80
KIN, NEF, NLC 1	3.	8.	1.					
$[\gamma]$ 2	0.	0.	0.	0.	0.	0.	0.	0.
3								
4	1.	1.	0.	0.	0.	1.	0.	1.
5								
6	0.	0.	0.	1.	0.	0.	0.	0.
7								
$[k]$ 8	8000.	4000.	-800.	0.	0.	0.	0.	0.
9	4000.	8000.	-800.	0.	0.	0.	0.	0.
10	-800.	-800.	1066.5	0.	0.	0.	0.	0.
11	0.	0.	0.	0.	9000.	0.	0.	0.
12	0.	0.	0.	18000.	0.	14400.	7200.	-865.
13	0.	0.	0.	0.	0.	7200.	14400.	-865.
14	0.	0.	0.	0.	0.	-865.	-865.	-69.1
$[P_6]$ 15	0.	0.	0.	133.33	-133.33	0.	0.	0.
16								
17								
18								
19								
20								

Figure 2-8

KTN= 3   NEFF= 8   NLC= 1   NJOIN= 0

DEFORMATION-DISPLACEMENT MATRIX--(GAMMA)

```
0.0000    1.0000    0.0000
0.0000    1.0000    0.0000
0.0000    0.0000    1.0000
0.0000    0.0000    1.0000
1.0000    0.0000    0.0000
1.0000    0.0000    0.0000
0.0000    1.0000    0.0000
0.0000    0.0000    1.0000
```

DEFORMATION-FORCE MATRIX--(K)

```
8000.0000    4000.0000   -800.0000     0.0000       0.0000       0.0000       0.0000       0.0000
4000.0000    8000.0000   -800.0000     0.0000       0.0000       0.0000       0.0000       0.0000
-800.0000   -800.0000   -106.5000     0.0000       0.0000       0.0000       0.0000       0.0000
0.0000       0.0000      0.0000    18000.0000    9000.0000    14400.0000    7200.0000   -865.0000
0.0000       0.0000      0.0000     9000.0000   18000.0000    7200.0000   14400.0000   -865.0000
0.0000       0.0000      0.0000        0.0000       0.0000   14400.0000    7200.0000   -865.0000
0.0000       0.0000      0.0000        0.0000       0.0000    7200.0000   14400.0000   -865.0000
0.0000       0.0000      0.0000        0.0000       0.0000    -865.0000   -865.0000    -69.1000
```

FIXED-END FORCES MATRIX

```
0.0000
0.0000
0.0000
-133.3300
-133.3300
0.0000
0.0000
0.0000
```

THE EXTERNAL STIFFNESS MATRIX--(S)

```
18000.0000      0.0000   18000.0000
   0.0000   38400.0000    -865.0000
```

58

```
18000.0000      -865.0000      18069.0900

THE REDUNDANT DISPLACEMENT MATRIX

      -0.0074
      -0.0000
      -0.0000

MEMBER-END FORCES DUE TO MEMBER-END DEFORMATIONS

      -0.0004
      -0.0004
       0.0001
    -133.3294
     -66.6647
       0.0003
       0.0006
      -0.0000

TOTAL MEMBER END FORCES

      -0.0004
      -0.0004
       0.0001
       0.0006
    -199.9947
       0.0003
       0.0006
      -0.0000
OBJECT CODE=    7768 BYTES,ARRAY AREA=    32400 BYTES,UNUSED=     9832 BYTES

HASP-II JOB W094 STATISTICS --        174 CARDS READ --        328 LINES PRINTED --        0 CARDS PUNCHED
                                  132K PARTITION SIZE --    132K CORE REQUESTED
                              00:00:09.92 TOTAL TIME
```

Figure 2-9

PROGRAM 1

```
        /WAT4,CEI40996,TIME=02,PAGES=020                        70.156       JOB 094
        LIMITS=(T=2,P=20)
        CLASS=W,PRIORITY=12,REGION=0132K,READER=READER1
C       PROGRAM FOR STIFFNESS OR DISPLACEMENT MATRIX APPROACH
C       NOTATIONS
C       KIN=NO. OF DEGREES OF KINEMATICALLY INDETERMINATE
C       NFF=NO. OF MEMBER DEFORMATIONS
C       NLC=NO. OF LOADING CONDITIONS
C       NJOIN= NO. OF TRANSVERSE JOINT FORCES
        DIMENSION GAMMA(30,10),C(30,30),GAMMAT(10,30),GAMMAC(30,3
       10),S(60,60),SV(30,30),Y(30,5),DI(30,5),CG(30,10),PI(30,5),P(30,5),
       2PJOIN(30,5)
      1 READ(1,10) JJ
        IF(JJ.LE.0) GO TO 20
        WRITE(3,30) JJ
     30 FORMAT(1H1,10X,'PROBLEM',I5 )
C       *****************
        READ(1,10) KIN,NFF,NLC,NJOIN
C       *****************
     10 FORMAT(4I5)
        WRITE(3,50) KIN,NFF,NLC,NJOIN
     50 FORMAT(1H1, 10X,'KIN=',I5,5X,'NFF=',I5,5X,'NLC=',I5,5X,'NJOIN=',
       1I5//)
C       *****************
        READ(1,100) ((GAMMA(I,J),I=1,NEF),J=1,KIN),((C(I,J),I=1,NEF),J=1,N
       1EF),((PO(I,J),I=1,NEF),J=1,NLC)
C       *****************
    100 FORMAT(8F10.4)
        DO 111 I=1,KIN
        DO 111 J=1,NLC
    111 PJOIN(I,J)=0.
        IF(NJOIN.EQ.0) GO TO 444
        READ(1,222) I,J,PIJ
    333 READ(1,222) I,J,PIJ
C       *****************
    222 FORMAT(2I5,F10.4)
        PJOIN(I,J)=PIJ+PJOIN(I,J)
        GO TO 333
    444 WRITE(3,110)
    110 FORMAT(///,' DEFORMATION-DISPLACEMENT MATRIX--(GAMMA)' //)
        DO 120 I=1,NEF
    120 WRITE(3,180) (GAMMA(I,J),J=1,KIN)
        WRITE(3,140)
    140 FORMAT(///,' DEFORMATION-FORCE MATRIX--(K)' //)
        DO 150 I=1,NEF
    150 WRITE(3,180) (C(I,J),J=1,NEF)
        WRITE(3,160)
    160 FORMAT(///,' FIXED-END FORCES MATRIX' //)
        DO 170 I=1,NEF
    170 WRITE(3,180) (PO(I,J),J=1,NLC)
    180 FORMAT(9F14.4/)
C       CALCULATION OF THE EXTERNAL STIFFNESS MATRIX--(S)
C       (S) IS EQUAL TO (GAMMAT)(K)(GAMMA)
        CALL TRANS(GAMMA,GAMMAT,KIN,NEF)
        CALL MULT(GAMMAT,C,GAMMAC,KIN,NEF,NEF)
        CALL MULT(GAMMAC,GAMMA,S,KIN,NEF,KIN)
```

```
      WRITE(3,190)
  190 FORMAT(////,' THE EXTERNAL STIFFNESS MATRIX--(S)' //)
      DO 210 I=1,KIN
  210 WRITE(3,180) (S(I,J),J=1,KIN)
C     INVERSION OF THE STIFFNESS MATRIX,(S)
      KIN2=KIN*2
      CALL VERSE(S,SV,KIN,KIN2)
      IF(KIN.EQ.1) SV(1,1)=1./S(1,1)
C     CALCULATE THE PRODUCT OF (GAMMAT) AND (PO)
      CALL MULT(GAMMAT,PO,Y,KIN,NEF,NLC)
      DO 400 J=1,KIN
  400 Y(I,J)=(-1.)*Y(I,J)
C     CALCULATE THE REDUNDANT DISPLACEMENT MATRIX
      CALL MULT(SV,Y,D1,KIN,KIN,NLC)
      WRITE(3,410)
  410 FORMAT(////,' THE REDUNDANT DISPLACEMENT MATRIX' //)
      DO 420 I=1,KIN
  420 WRITE(3,180) (D1(I,J),J=1,NLC)
C     CALCULATE THE (P1) FORCES,(P1)=(K) (GAMMA) (D1)
      CALL MULT(C,GAMMA,CG,NEF,KIN,KIN)
      CALL MULT(CG,D1,P1,NEF,KIN,NLC)
      WRITE(3,510)
  510 FORMAT(////,' MEMBER-END FORCES DUE TO MEMBER-END DEFORMATIONS' //)
      DO 520 I=1,NEF
  520 WRITE(3,180) (P1(I,J),J=1,NLC)
C     CALCULATE THE TOTAL MEMBER END FORCES, (P)=(NP1)+(PO)
      CALL ADD(P0,P1,P,NEF,NLC)
      WRITE(3,430)
  430 FORMAT(////,' TOTAL MEMBER END FORCES' //)
      DO 440 I=1,NEF
  440 WRITE(3,180) (P(I,J),J=1,NLC)
      GO TO 1
   20 STOP
      END
      SUBROUTINE TRANS(U,V,K,L)
C     THIS SUBROUTINE WILL TRANSPOSE A MATRIX
      DIMENSION U(L,K),V(K,L)
      DO 10 I=1,K
      DO 10 J=1,L
   10 V(I,J)=U(J,I)
      RETURN
      END
      SUBROUTINE MULT(U,V,T,M,K,N)
C     THIS WILL MULTIPLY TWO MATRICES
      DIMENSION U(M,K),V(K,N),T(M,N)
      DO 20 I=1,M
      DO 20 J=1,N
      T(I,J)=0.
      DO 20 L=1,K
      T1=U(I,L)*V(L,J)
   20 T(I,J)=T(I,J)+T1
      RETURN
      END
```

PROGRAM I (cont.)

```
85          SUBROUTINE VERSE(U,V,M,L,N)
86    C     THIS WILL INVERSE A GIVEN MATRIX BY GAUSS-JORDAN METHOD
87          DIMENSION U(L,N),V(M,M)
88          M1=M+1
89          DO 32 I=1,M
90          DO 32 J=M1,N
91          JI=J-I
92          IF(JI-M) 36,33,36
93    36    U(I,J)=0.
94          GO TO 32
95    33    U(I,J)=1.
96    32    CONTINUE
97          K1=K+1
98          K3=K+M-1
99          IF(U(K,K).NE.0.) GO TO 60
100   59    IF(U(I1,K).NE.0.) GO TO 58
101         I1=I1+1
102         GO TO 59
103   58    DO 50 J=1,N
104   50    U(K,J)=U(K,J)+U(I1,J)
105   60    DO 51 J=1,N
106   51    U(K+M,J)=U(K,J)/U(K,K)
107         DO 54 I=K1,K3
108         R=U(I,K)
109         DO 54 J=1,N
110   54    U(I,J)=U(I,J)-U(K+M,J)*R
111         DO 55 I=1,M
112         DO 55 J=1,M
113   55    V(I,J)=U(I+M,J+M)
114         RETURN
115         END
116         SUBROUTINE ADD(U,V,T,K,L)
117         DIMENSION U(K,L),V(K,L),T(K,L)
118         DO 70 I=1,K
119         DO 70 J=1,L
120   70    T(I,J)=U(I,J)+V(I,J)
121         RETURN
122         END
123   /DATA
```

4 The matrix operations involved to obtain the displacement matrix $\{\Delta\}$ and member-forces matrix $\{P\}$ are given by Equations (2-55) and (2-57). These operations are done by the digital computer using the program at the end of Chapter 2. The results of the computer analysis are given in Figs. 2-8 and 2-9.

COMPUTER PROGRAM FOR THE ANALYSIS OF TYPICAL STRUCTURES

The computer Program I is written in FORTRAN IV language and has 123 statements. It is a simple program, but it is an efficient one to analyze continuous beams and rigid frames. It will handle various loading conditions such as concentrated loads or uniform loads and their combinations. The input matrices are read in by rows. INPUT quantities are

    KIN = Degree of kinematic indeterminacy
    NEF = Number of member deformations
    NLC = Number of loading conditions
    [R] = Deformation-displacement matrix
    [K] = Deformation-force matrix
    $[P_0]$ = Fixed-end forces

The OUTPUT quantities are

    [S] = External stiffness matrix of the structure
    [Δ] = Redundant displacement matrix
    [P] = Total member-end forces

To illustrate the use of this computer program, several problems are solved by it. Its applications are extended to continuous beams as well as to rigid frames as shown in this chapter. The INPUT data are prepared correctly and completely on actual data forms with exact INPUT FORMATS. The student, studying this information will be able to apply the program to many various structural analysis problems. Let us remember that the finite element method will use the stiffness matrix analysis in its development.

PROBLEMS

2-1 Perform the structural analysis of the given plane truss using the force matrix method. Generate the necessary matrices and use computers to obtain solutions when practical. The members are made of A-36 steel. The geometry and loading of the truss is shown in Fig. 2-10. All members have a constant AE/L value.

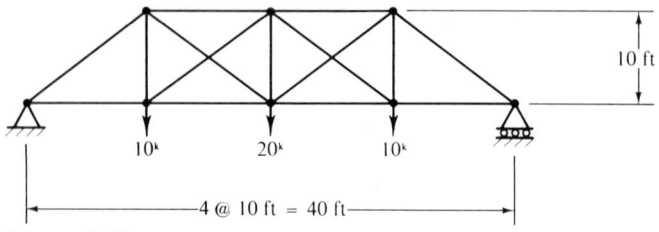

Figure 2-10

2-2 Analyze the rigid frame shown in Figure 2-6 using the displacement matrix method. Generate the necessary matrix quantities and when practical use a computer to perform matrix operations. The frame is made of A-36 steel. The geometry, loading, and boundary conditions are given in Fig. 2-11. Draw the final shear and moment diagrams.

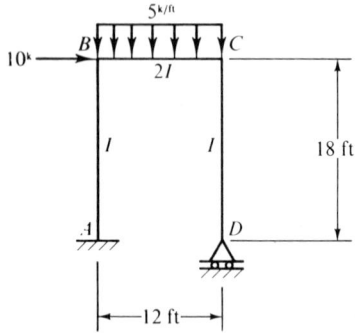

Figure 2-11

2-3 Repeat the analysis of the rigid frame of Problem 2-2 using the force matrix method. Compare the results of the two analysis.

2-4 Perform the stiffness matrix method analysis of the given continuous beam and draw the final shear and moment diagrams. Young's modulus is 30,000 ksi. The geometry, boundary conditions, and loads are shown in Fig. 2-12.

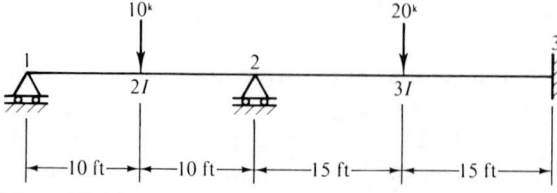

Figure 2-12

2-5 Perform the stiffness matrix method analysis of the given two-bay-frame, Fig. 2-13. The frame is built using A-36 steel. Draw the final bending moment diagram.

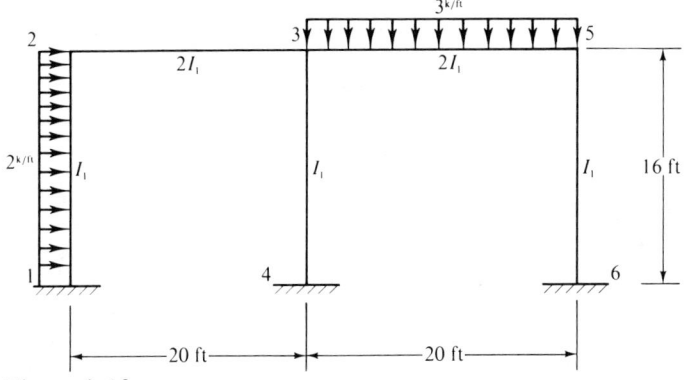

Figure 2-13

REFERENCES

1 C. K. WANG, *Matrix Methods of Structural Analysis.* New York and London: Intext Educational Publishers, 1970. p. 62.
2 S. T. CARPENTER, *Structural Mechanics.* New York: Wiley, 1960, pp. 95–104.
3 J. I. PARCEL and R. B. B. MOORMAN, *Analysis of Statically Indeterminate Analysis.* New York: Wiley, 1955, pp. 206–214.
4 JAMES M. GERE and W. WEAVER, JR., *Analysis of Framed Structures.* Princeton, N.J.: Van Nostrand, 1965.
5 H. I. LAURSEN, *Matrix Analysis of Structures.* New York: McGraw-Hill, 1966.
6 J. S. PRZEMIEMECKI, *Theory of Matrix Structural Analysis.* New York: McGraw-Hill, 1968.
7 M. F. RUBENSTEIN, *Matrix Computer Analysis of Structures.* Englewood Cliffs, N.J.: Prentice-Hall, 1966.
8 H. C. MARTIN, *Introduction to Matrix Methods of Structural Analysis.* New York: McGraw-Hill, 1966.
9 OKTAY URAL, *Matrix Operations and Use of Computers in Structural Engineering.* New York and London: Intext Educational Publishers, 1971.

The Finite Element Method in Structural Analysis

3

INTRODUCTION

Recent technological advancements have created needs for accurate analyses of very large and complex structural systems. Geometric irregularities, varying material properties, and multiple loading conditions associated with analysis problems have introduced difficulties which begin to exceed the limitations of rigorous mathematical methods. The development of the high-speed electronic digital computers in the last twenty years has generated great interest in numerical methods related to the solution of engineering problems. At the same time, matrix concepts and operations have forced themselves into the formulation of many complex engineering analysis problems, since they have been very adaptable to the logical operations of digital computers. Accordingly, early in 1950 the engineers and scientists had for their innovative studies two great assets: matrix concepts and digital computers. They were able tools to formulate problems and then solve them quickly and correctly. Potentially, these two items relieved the engineer and the scientist for long hours of tedious mathematical computations, giving them additional precious time for creativity in their work.

Expanded knowledge of matrix structural analysis methods also contributed to the success of the finite element method, which has a physical approach to its definition. Many structures such as trusses and frames are formed by members interconnected at definite nodes. These members actually are one-dimensional finite elements which when connected at the nodes generate the whole structure. A general definition of the finite-element method would be an analysis method of a structural system which is represented as an assemblage of a finite number of discrete elements interconnected at a finite number of node points. When this concept is applied to an elastic continuum, the number of possible elements and connections among them becomes infinite. There is thus a need for an idealization of the continuum into a structural system with finite discrete elements. This approach and approximation have been shown to be acceptable [1]. Once the structural system is described by a mesh of finite elements, approximations are made on the behavior of these elements in order to approximate the behavior of the continuous structural system.

The finite-difference method which has been in practice for many years, is based on the use of differential equations whose solutions are approximated at equally separated points. The differential equations generated by the finite-difference method describe the behavior of the complete structural system. Therefore, the finite-difference method discretizes the differential equations to solve them.

The finite element method divides the structure into discrete elements and then generates equations for each element to express their behavior.

HISTORICAL SURVEY

Since 1940 there has been a growing amount of technical publications related to the finite element method. The initial concepts of the applications of this method were introduced by Hrennikoff [2], McHenry [3], and Grinter [4] when they used a system of one-dimensional elements to solve two-dimensional systems. The contributions of Argyris [5, 6] have been very valuable in the initial stages of the development of finite-element method. He presented the general approach to the computer analysis of structural systems. Turner, Clough, Martin, and Topp [1] presented the use of the two-dimensional plate element in the analysis of complex structures. Clough [7] introduced both triangular and rectangular plate elements for the analysis of plane stress elasticity problems. Wilson [8] prepared a computer program for solution of plane elasticity problems, and other programs followed.

The applicability of the finite element method to plate-bending problems has been shown. Melosh [9] developed the stiffness matrix for an isotropic rectangular plate panel of uniform thickness and also discussed the basis for derivation of matrices for the direct stiffness and method [10]. Clough and Rashid [11] demonstrated the application of the finite element method to axially symmetric elastic solids. They developed special stiffness matrices for the axisymmetric finite element. Wilson [12] presented an approach for the determination of stresses and displacements in axisymmetric solids of arbitrary geometry subjected to thermal and mechanical loads. Tocher and Hartz [13] discussed a higher-order finite element for plane stress analysis.

The analysis of three-dimensional solids has been studied by several authors [14-16]. Tong and Pian [17] presented the theoretical development of the conditions to insure the finite-element displacement analysis to converge to the exact displacements solutions when the element sizes are reduced. Meantime, there were many publications for the improvement of existing concepts and their applications. Akyüz and Utku [18] developed an automatic node labeling scheme. As the knowledge of the method expanded, its applications began to invade the field of shell analysis. Adini [19] discussed the analysis of shell structures by the finite element method. The contributions of Utku [20] to this area are valuable. Zienkiewicz [21], Meek [22], and Ural [23] have discussed the finite element method and their applications to structural analysis problems.

It is an almost impossible task to locate and list all the publications related to the finite element method. There are many of them dealing with theoretical developments and applications. The ones which are referenced here are selected to give a general introduction to the historical background of the evolution of this method, which is finding more applications every passing day.

THEORETICAL DISCUSSION

The concept of the finite element method is the idealization of a continuous media by an articulated structural system to which the matrix methods of structural analysis are directly applicable. There are two matrix method approaches associated with the finite element method [1, 23].

    a The force finite element method assumes the internal forces as the unknowns of the problem. To generate the governing equations, first the equilibrium conditions are used; then to develop the additional equations which might be necessary to obtain a solution, compatibility conditions are introduced.

b The displacement finite element method assumes the displacements of the nodes as the unknowns of the problem. In this approach the compatibility conditions in and among elements are initially satisfied. Then the governing equations in terms of node displacements are written for each nodal point using the equilibrium conditions. The theoretical justification of the displacement finite element method can be shown by the use of the principle of minimum potential energy.

The differences between the force and displacement finite element method lie on the selection of the unknowns of the analysis and the variations in the matrix quantities associated with their formulations. The experience has shown that the displacement finite element method is more desirable since its formulation is simpler for the majority of structural analysis problems. Therefore, this book will use the displacement finite element method for the treatment of the topic and for the analysis of structural systems.

Matrix structural analysis methods have five major steps to perform to obtain a solution for the given problem. They are

1. Idealization of the structural system.
2. Definition of the geometric and elastic properties of the system.
3. Description of the boundary conditions.
4. Definition of the loading conditions.
5. Operations on the generated matrix quantities to obtain the displacements and stresses.

The development of the most of these matrix quantities can be automated by the use of relevant computer programs. The development of element stiffnesses is based on the principles of continuum mechanics. Matrix theory is simply used to combine and solve the resulting equations.

It might be enlightening to the reader at this time to present the routine followed to apply the finite element method to a problem. Each part of this routine will be fully explored and discussed later in the book.

1. Idealization of the structural system. This requires the selection of the type and size of the finite elements to generate the mesh of the system.
2. Generation of the stiffness matrix quantities for the elements.
3. Superposition of the element stiffness matrices to develop the stiffness matrix of the total structural system.

4 Determination of the unknown nodal displacements of the problem by the solution of the system of linear equations obtained using the equilibrium conditions at the nodes.

5 Computations of all other required values such as stresses and strains associated with the problem.

Usually the accuracy and effectiveness of the finite element method will depend on the type and number of elements used in the mesh generation. The decision on the types of elements to be selected depends on the geometry and nature of the problem. The most widely used element is the constant-strain triangular finite element. This element is desirable since it molds into various types of boundary conditions and the theory associated with it is comparatively simple. Complex analysis problems might use other types of finite elements which are also readily available. Some of them are line segment, triangle, quadrilateral, conical segment, tetrahedron, hexahedron, triangular torus, quadrilateral torus, curved triangle, and curved rectangle. The selection of a particular type of element or any combination of elements will greatly depend on the knowledge and competence of the engineer. In the theoretical discussion of the finite element method, the concepts of nodal displacements, nodal forces, and element stiffness matrices are used. In the analysis of linear elastic structures, the above-mentioned terms might be thought as analogous to joint displacements, joint forces, and member stiffness matrices. Another concern of the analyst is to secure compatibility at the boundaries of adjoining elements. This can be done by the assumption of proper general displacement functions for the elements.

The applications of the displacement finite element method to many areas of engineering such as the analysis of linear structures, plates, shells, solids, dams, slope-stability, foundations, fluid flow, dynamics, stress concentrations, tunnels, and rock mechanics are taking place with success.

ENERGY PRINCIPLE

All the finite element method relations which are used for the solution of problems need to obey the energy principle, directly or indirectly. It must be observed that the sum of the product of all displacements and corresponding load components should be equal to the sum of the product of all strains and corresponding stresses. They must represent the true external and internal works related to the structural system. An idealization is necessary, since the distributed loads at the element boundaries are replaced by equivalent loads applied at the nodes.

The total potential energy of the system might include four distinct contributions.

$$(V) = (U) - (U_F) - (U_R) - (U_S) \tag{3-1}$$

where V is the total potential energy, U the internal strain energy, U_F the potential energy due to body forces, U_R the potential energy due to nodal forces, and U_S the potential energy due to surface traction.

If the structural system considered in the analysis enjoys full internal and external equilibrium and if a small variation in the displacements occurs, then the introduced additional external work must be equal to the resulting internal work. This leads to the fact that the variation of the total potential energy $d(V)$ must be equal to zero:

$$d(V) = d[(U) - (U_F) - (U_R) - (U_S)] = 0 \tag{3-2}$$

Equation (3-2) denotes that the total potential energy of the system must have a stationary value and, for the equilibrium condition, this value is to be a minimum. The displacement finite element analysis obeys the principle of minimum potential energy, which states that of all displacements expressions that satisfy the displacement boundary conditions, the one satisfying equilibrium condition makes the total potential energy a minimum.

Since the potential energy of the system, V, will be a function of the unknown nodal displacements, the partial derivatives of this stationary quantity with respect to the unknowns will vanish:

$$\frac{\partial V}{\partial \delta_1} = \frac{\partial V}{\partial \delta_2} = \frac{\partial V}{\partial \delta_3} = \cdots = \frac{\partial V}{\partial \delta_n} = 0 \tag{3-3}$$

Expression (3-3) will generate the basic equilibrium equations, (3-14), which are necessary to solve for the unknown displacements. This energy discussion will also develop an expression for the element stiffness matrix which will be presented with details in Chapter 4.

CONVERGENCE [16, 26]

The validity of the finite element method greatly depends on how well the approximate solution converges to the exact solution. It has been shown that convergence will occur if three general conditions are satisfied.

1. The displacements must be continuous within an element and displacement compatibility along the boundaries of adjacent elements must be secured. The first part of the above statement can

be secured by the selection of continuous polynomials as the general displacement functions. The second part does prohibit random and independent displacement of adjacent elements at their adjoining boundaries.

2. The assumed displacement fields must be capable of representing a state of constant strain. The necessity of this requirement can be physically explained if one conceives an infinitesimal element whose strain will approach a constant value.

3. The assumed displacement fields must include rigid body motion. This will allow all the points of an element to go through similar displacements.

As the problem of analysis becomes complex, the satisfaction of the above conditions might require careful consideration of the nature of the assumed displacement functions and the type and size of the finite elements.

ADAPTABILITY [10]

The finite element analysis is based on the stationary total potential energy principle, using a variation of Ritz procedure. The difference between the finite element and Ritz methods for the solution of stationary value problems lies in the type of the functions used in their operations. The Ritz method uses smooth displacement functions to describe the behavior of the entire structural system whereas the finite element method uses displacement functions for each element of the system. This sound theoretical background of the finite element method secures the initial requirement for its adaptability to actual problems. One of its greatest assets is to describe a continuous media by discrete, interconnected finite elements of various shapes. Hence a two-and three-dimensional structural continuum can be idealized into an articulated system to which standard structural analysis methods are applicable.

The adaptability of the finite element method should not be taken for granted, even though it is a very powerful approach for the formulation of complex structures. The geometric approximation of a continuous surface by elements must not deviate greatly from the true shape. The displacement approximations must be consistent with the geometric approximations. A simple triangular constant strain element should have a simple displacement function.

The use of the finite element method has already invaded the analysis of problems of various characteristics. Programs based on this method are readily available for the solution of one, two, and three dimensional static and dynamic analysis problems. These computer programs will be discussed in Chapters 8 and 9.

GENERAL FORMULATION OF THE FINITE ELEMENT METHOD

The general steps necessary to perform a finite element analysis which were previously mentioned, need to be mathematically formulated and discussed. This will be a transition from matrix methods of structural analysis presented in Chapter 2 to the finite element method. A plane stress problem is considered using the constant-strain triangular element. Let us consider a thin plate in pure tension located in X-Y coordinate plane, Fig. 3-1. The plate is divided into triangular elements connected at

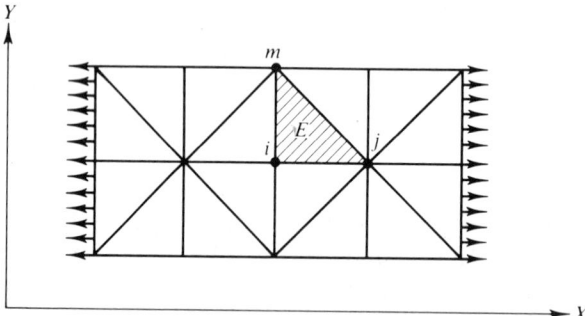

Figure 3-1

the nodes. This could be called a *coarse-mesh generation* for illustrative purposes. Due to the nature of the problem, the nodes have only two degrees of freedom, namely in the X and Y directions.

    a The displacements of the nodes are the unknowns of the displacement finite element method.

$$\{\delta\}_E = \begin{Bmatrix} \delta_i \\ \delta_j \\ \delta_m \end{Bmatrix} \qquad (3\text{-}4)$$

The subscript E defines the element E. The subscripts i, j and m denote the nodes associated with the element E. Each node will have its two degrees of freedom.

$$\{\delta_i\} = \begin{Bmatrix} u_i \\ v_i \end{Bmatrix} \qquad (3\text{-}5)$$

GENERAL FORMULATION OF THE FINITE ELEMENT METHOD   75

The $u(x,y)$ is the displacement in the X direction and $v(x,y)$ is the displacement in the Y direction. When Equation (3-5) is properly substituted in Equation (3-4), a complete node-displacement matrix is generated.

b To describe the behavior of the elements, general displacement expressions as functions of nodal displacements are assumed.

$$\{\phi\} = \begin{Bmatrix} u(x,y) \\ v(x,y) \end{Bmatrix} \quad (3\text{-}6)$$

The terms $u(x,y)$ and $v(x,y)$ describe the displacements at any interior point of the element. The displacement function's nature should be compatible with the type of the elements used. In the case of a constant strain triangular element, the displacement functions are linear. These functions should also satisfy the displacement compatibility at the boundaries of the adjoining elements. The unknown coefficients of the assumed functions are determined by matrix operations which will be discussed in detail in Chapter 4.

c The determination of the nodal displacements will allow the computations of all the necessary strain values in the elements. Each point will have three strain components.

$$\{\epsilon\} = \begin{Bmatrix} \epsilon_x \\ \epsilon_y \\ \gamma_{xy} \end{Bmatrix} \quad (3\text{-}7)$$

The theory of elasticity [25] defines the relation between displacements and the resulting strains.

$$\{\epsilon\} = \begin{Bmatrix} \dfrac{\partial u}{\partial x} \\ \dfrac{\partial v}{\partial y} \\ \dfrac{\partial u}{\partial y} + \dfrac{\partial v}{\partial x} \end{Bmatrix} \quad (3\text{-}8)$$

Equation (3-8) clearly illustrates the reason why a constant-strain triangular element should have a linear displacement function associated with it. A general formulation of the displacement-strain relation is given by the matrix equation

$$\{\epsilon\} = \{B\}\{\delta\} \quad (3\text{-}9)$$

d The obtained strain values will allow the calculations of the stresses in the elements. The theory of elasticity relates strains to stresses by a matrix $[D]$. There are three stress components at any interior point of an element.

$$\{\sigma\} = \begin{Bmatrix} \sigma_x \\ \sigma_y \\ \tau_{xy} \end{Bmatrix} \tag{3-10}$$

Using the proper elasticity transformation matrix, the following relation is written:

$$\{\sigma\} = [D]\{\epsilon\} \tag{3-11}$$

Equation (3-11) when expanded will have more illustrative form:

$$\begin{Bmatrix} \sigma_x \\ \sigma_y \\ \tau_{xy} \end{Bmatrix} = \frac{E}{1-\nu^2} \begin{bmatrix} 1 & \nu & 0 \\ \nu & 1 & 0 \\ 0 & 0 & \frac{1-\nu}{2} \end{bmatrix} \begin{Bmatrix} \epsilon_x \\ \epsilon_y \\ \gamma_{xy} \end{Bmatrix} \tag{3-12}$$

e In the case of the requirement of principal stresses, Equation (3-13) is needed for the computations.

$$\sigma_{max} = \frac{\sigma_x + \sigma_y}{2} + \sqrt{\left(\frac{\sigma_x - \sigma_y}{2}\right)^2 + \tau_{xy}^2}$$

$$\sigma_{min} = \frac{\sigma_x + \sigma_y}{2} - \sqrt{\left(\frac{\sigma_x - \sigma_y}{2}\right)^2 + \tau_{xy}^2} \tag{3-13}$$

The solution of the finite element method depends on the accurate determination of the unknown nodal displacements. The basic equation which yields their solution is the one used in the stiffness matrix method of analysis.

$$\{Q\} = [K]\{\delta\} \tag{3-14}$$

where $\{Q\}$ is the equivalent load matrix which is obtained by lumping the edge and element loads at the nodes, $\{\delta\}$ is the unknown nodal displacements matrix, and $[K]$ is the total stiffness matrix of the structural system. It is obtained by the superposition of the individual element stiffness matrices. The rearrangement of Equation (3-14) will directly give the displacements values.

$$\{\delta\} = [K]^{-1}\{Q\} \tag{3-15}$$

The generation of the element stiffness matrices and then the total stiffness matrix will be discussed to detail in Chapter 4.

REFERENCES

1. J. J. TURNER, R. W. CLOUGH, H. C. MARTIN, and L. J. TOPP, "Stiffness and Deflection Analysis of Complex Structures," *Journal of Aeronautical Sciences* **23**, 805–832 (1956).
2. A. HRENNIKOFF, "Solution of Problems in Elasticity by the Frame Work Method," *Journal of Applied Mechanics* **8**, 4, (December 1941).
3. D. MCHENRY, "A Lattice Analogy for the Solution of Stress Problems," *Journal of the Institution of Civil Engineers*, 59–82, (December 1943).
4. L. E. GRINTER, "Statistical State of Stress Studied by Grid Analysis," Numerical Methods of Analysis," Numerical Methods of Analysis in L. E. Grinter, *Engineering*. New York: Macmillan, 1949.
5. J. H. ARGYRIS, "Energy Theorems and Structural Analysis," *Aircraft Engineering* **26–27** (October 1954–May 1955).
6. J. H. ARGYRIS, "The Matrix Theory of Statics," *Ingenieur-Archiv*, **25**, 174, (1957).
7. R. W. CLOUGH, "The Finite Element Method in Plane Stress Analysis," *Proceedings, ASCE*, 2nd Conference on Electronic Computation, Pittsburgh, Pa., 345–378, (September 1960).
8. E. L. WILSON, *Structural Engineering Laboratory Report 63-2*, University of California-Berkeley, 1963.
9. R. J. MELOSH, "A Stiffness Matrix for the Analysis of Thin Plates in Bending," *Journal of the Aerospace Sciences*, 34–41, (January 1961).
10. R. J. MELOSH, "Basis for Derivation of Matrices for the Direct Stiffness Method," *AIAA Journal* **1**, 7, 1631–1637 (July 1963).
11. R. W. CLOUGH and Y. RASHID, "Finite Element Analysis of Axi-Symmetric Solids," *Journal of the Engineering Mechanics Division*, Proc. of the American Society of Civil Engineers, 71–85, (February 1965).
12. E. L. WILSON, "Structural Analysis of Axisymmetric Solids," *AIAA Journal* **3**, 12, 2269–2274 (December 1965).
13. J. L. TOCHER and B. J. HARTZ, "Higher-Order Finite Element for Plane Stress," *Journal of Engineering Mechanics Division, Proceedings of the ASCE*, 149–172 (August 1967).
14. R. H. GALLAGHER, J. PADLOG, and P. P. BIJLAARD, "Stress Analysis of Heated Complex Shapes," *ARS Journal* **32**, 700–707 (1962).
15. H. C. MARTIN, "Plane Elasticity Problems and the Direct Stiffness Method," *The Trend in Engineering* **13**, January 1961.
16. R. J. MELOSH, "Structural Analysis of Solids," *Journal of the Structural Division, Proceedings of the ASCE*, 205–223 (August 1963).
17. PIN TONG and T. H. H. PIAN, "The Convergence of Finite Element Method

in Solving Linear Elastic Problems," *Journal of Solids Structures,* **3,** 865–879 (1967).

18 F. A. AKYÜZ and S. UTKU, "An Automatic Node Relabelling Scheme for Band-width Minimization of Stiffness Matrices," *AIAA Journal* **6,** 4, 728–730 (April 1968).

19 A. ADINI, "Analysis of Shell Structures by the Finite Element Method," Ph.D. thesis, University of California, Berkeley, 1961.

20 S. UTKU, "Stiffness Matrices for Thin Triangular Elements of Non-Zero Gaussian Curvature," *AIAA Journal* **5,** 9 (September 1967).

21 O. C. ZIENKIEWICZ and Y. K. CHEUNG, *The Finite Element Method in Structural and Continuum Mechanics.* New York and London: McGraw-Hill, 1967.

22 J. L. MEEK, *Matrix Structural Analysis.* New York: McGraw-Hill, 1971.

23 OKTAY URAL, *Matrix Methods and Use of Computers in Structural Engineering.* New York and London: Intext Educational Publishers, 1971.

24 B. FRAEIJS DE VENBEKE, "Duality between Displacement and Equilibrium Method with a View to Obtaining Upper and Lower Bounds to Static Influence Coefficients," *Proceedings,* 14th Meeting of the AGARD Structures and Materials Panel, Paris, July 1962.

25 S. TIMOSHENKO and J. H. GOODIER, *Theory of Elasticity,* 3rd ed. New York: McGraw-Hill, 1970.

26 E. R. OLIVEIRA, "Theoretical Foundation of the Finite Element Method," *International Journal of Solids and Structures* **4,** (1968).

Plane Stress and Plane Strain Analysis

4

INTRODUCTION

The finite element method idealizes a plane stress system by a set of connected elements to which the stiffness matrix method of structural analysis can be directly applied. Actually, the analysis of two-dimensional elasticity problems were the first successful application of the method [1]. Triangular elements have been widely used because of their ability to define boundaries with relative ease and because expressions related to them are comparatively simple.

Continuity between the constant strain triangular elements is secured by the fact that the strains in the elements are constant. This will allow only straight-line displacements. The loads acting on the structure are replaced by statically equivalent concentrated loads applied at the nodes of the idealized structure.

In the plane stress case [2], the state of stress is defined only by the stress components σ_x, σ_y, and τ_{xy}, in the x-y-z coordinates system. The analysis of a thin plate under uniform tensile load is performed as a plane stress problem.

In the plane strain case, the state of strain in the x-y-z coordinates system is defined by the strain components ϵ_x, ϵ_y, and τ_{xy}.

Plane stress and plane strain analyses follow similar logic in their solution. The only variation will be on the elasticity matrix $[D]$ which relates the strains and stresses. The presentation in this chapter will be the analysis of a two dimensional elasticity problem using constant strain triangular elements.

THEORETICAL DEVELOPMENT

The constant strain triangle which will be the case of our discussion is shown in Fig. 4-1.

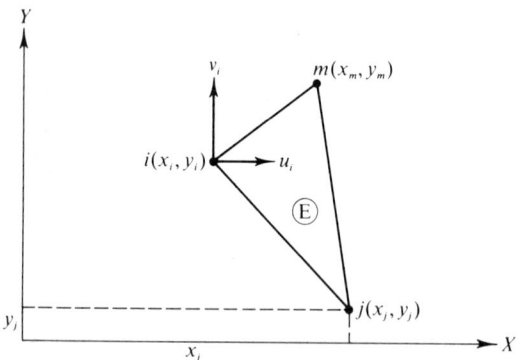

Figure 4-1

    a The unknowns of the analysis are the nodal displacements $\{\delta\}$. Each node point will have two degrees of freedom, namely, one in the X direction and the other in the Y direction. The nodal displacement matrix will have six elements, since there are three nodes and each node has two degrees of freedom.

$$\{\delta\} = \begin{Bmatrix} \delta_i \\ \delta_j \\ \delta_m \end{Bmatrix} = \begin{Bmatrix} u_i \\ v_i \\ u_j \\ v_j \\ u_m \\ v_m \end{Bmatrix} \quad (4\text{-}1)$$

The X displacements are expressed by the letter u and the Y displacements are expressed by the letter v.

b The finite element method requires the assumption of a general displacement expression as a function of the unknown nodal displacements. It has to describe the displacements of all the points of the element. The same general function should also satisfy the displacement compatibility conditions at the boundaries of the elements. In the case of the constant-strain triangular element, the general displacement function has to be a linear one—for example,

$$\{\phi\} = \begin{Bmatrix} u(x,y) \\ v(x,y) \end{Bmatrix} = \begin{Bmatrix} a_1 + a_2 x + a_3 y \\ a_4 + a_5 x + a_6 y \end{Bmatrix} \tag{4-2}$$

This requirement is clearly denoted by Equation (3-8). Since the strains are equal to the partial derivatives of the displacement functions, and since constant-strain triangles are used, the displacement functions $u(x,y)$ and $v(x,y)$ have to be linear in nature.

In Equation (4-2), $\{\phi\}$ is the general displacement function; $u(x,y)$ is any X displacement, and $v(x,y)$ is any Y displacement in the element. $a_1, a_2, a_3, \ldots, a_6$ are the undetermined coefficients of the equations. Expression (4-2) can be rewritten in the form

$$\{\phi\} = \begin{Bmatrix} a_1 + a_2 x + a_3 y \\ a_4 + a_5 x + a_6 y \end{Bmatrix} = \begin{bmatrix} 1 & x & y & 0 & 0 & 0 \\ 0 & 0 & 0 & 1 & x & y \end{bmatrix} \begin{Bmatrix} a_1 \\ a_2 \\ a_3 \\ a_4 \\ a_5 \\ a_6 \end{Bmatrix} \tag{4-3}$$

To determine the constants $a_1, a_2, a_3, \ldots, a_6$, the coordinates of the nodal points are substituted in the general displacement function (4-2). This will generate six linear equations in which the only unknowns are the coefficients. For X displacements, the three generated linear equations are

$$\begin{aligned} u_i &= a_1 + a_2 x_i + a_3 y_i \\ u_j &= a_1 + a_2 x_j + a_3 y_j \\ u_m &= a_1 + a_2 x_m + a_3 y_m \end{aligned} \tag{4-4}$$

For Y displacements, the three obtained equations are

$$\begin{aligned} v_i &= a_4 + a_5 x_i + a_6 y_i \\ v_j &= a_4 + a_5 x_j + a_6 y_j \\ v_m &= a_4 + a_5 x_m + a_6 y_m \end{aligned} \tag{4-5}$$

The solution of the system of linear equations (4-4) is shown for illustrative purposes, using matrix notation and matrix operations:

$$\{u\} = \{1xy\}\{a\} \qquad (4\text{-}6)$$

Equation (4-6), which is a concise notation for Equation (4-4), is rewritten as

$$\begin{Bmatrix} u_i \\ u_j \\ u_m \end{Bmatrix} = \begin{bmatrix} 1 & x_i & y_i \\ 1 & x_j & y_j \\ 1 & x_m & y_m \end{bmatrix} \begin{Bmatrix} a_1 \\ a_2 \\ a_3 \end{Bmatrix} \qquad (4\text{-}7)$$

Equation (4-7) is rearranged to solve for the coefficients a_1, a_2, and a_3.

$$\begin{Bmatrix} a_1 \\ a_2 \\ a_3 \end{Bmatrix} = \begin{bmatrix} 1 & x_i & y_i \\ 1 & x_j & y_j \\ 1 & x_m & y_m \end{bmatrix}^{-1} \begin{Bmatrix} u_i \\ u_j \\ u_m \end{Bmatrix} \qquad (4\text{-}8)$$

The inverse operation of the above matrix may be done by the adjoint method using Equation (1-9).

$$\begin{bmatrix} 1 & x_i & y_i \\ 1 & x_j & y_j \\ 1 & x_m & y_m \end{bmatrix}^{-1} = \frac{\begin{bmatrix} c_{11} & c_{12} & c_{13} \\ c_{21} & c_{22} & c_{23} \\ c_{31} & c_{32} & c_{33} \end{bmatrix}^T}{\begin{vmatrix} 1 & x_i & y_i \\ 1 & x_j & y_j \\ 1 & x_m & y_m \end{vmatrix}} \qquad (4\text{-}9)$$

The elements of the numerator matrix of Equation (4-9) are defined as follows:

$$\begin{aligned} c_{11} &= x_j y_m - y_j x_m = \alpha_i \\ c_{21} &= y_i x_m - x_i y_m = \alpha_j \\ c_{31} &= x_i y_j - y_i x_j = \alpha_m \\ c_{12} &= y_j - y_m = \beta_i \\ c_{22} &= y_m - y_i = \beta_j \\ c_{32} &= y_i - y_j = \beta_m \\ c_{13} &= x_m - x_j = \gamma_i \\ c_{23} &= x_i - x_m = \gamma_j \\ c_{33} &= x_j - x_i = \gamma_m \end{aligned} \qquad (4\text{-}10)$$

The value of the determinant of Equation (4-9) is obtained by proper expansion using expression (1-7).

$$\begin{vmatrix} 1 & x_i & y_i \\ 1 & x_j & y_j \\ 1 & x_m & y_m \end{vmatrix} = x_i(y_j - y_m) + x_j(y_m - y_i) + x_m(y_i - y_j) \quad (4\text{-}11)$$

Mathematically, the value of expression (4-11) is equal to twice the area of the triangular element whose nodes have as coordinates, (x_i, y_i), (x_j, y_j), and (x_m, y_m).

$$\begin{bmatrix} 1 & x_i & y_i \\ 1 & x_j & y_j \\ 1 & x_m & y_m \end{bmatrix} = 2A \quad (4\text{-}12)$$

Using values of Equations (4-10) and (4-12), expression (4-8) is rewritten:

$$\begin{Bmatrix} a_1 \\ a_2 \\ a_3 \end{Bmatrix} = \frac{1}{2A} \begin{bmatrix} \alpha_i & \alpha_j & \alpha_m \\ \beta_i & \beta_j & \beta_m \\ \gamma_i & \gamma_j & \gamma_m \end{bmatrix} \begin{Bmatrix} u_i \\ u_j \\ u_m \end{Bmatrix} \quad (4\text{-}13)$$

Proceeding in the same manner, the values of the other coefficients are obtained.

$$\begin{Bmatrix} a_4 \\ a_5 \\ a_6 \end{Bmatrix} = \frac{1}{2A} \begin{bmatrix} \alpha_i & \alpha_j & \alpha_m \\ \beta_i & \beta_j & \beta_m \\ \gamma_i & \gamma_j & \gamma_m \end{bmatrix} \begin{Bmatrix} v_i \\ v_j \\ v_m \end{Bmatrix} \quad (4\text{-}14)$$

The general displacement function $\{\phi\}$ is determined by substituting expressions (4-13) and (4-14) in Equation (4-3). For illustration, let us derive the general x-displacement function $u(x, y)$ of $\{\phi\}$. Then the y-displacement function $v(x, y)$ can be readily obtained.

$$\{u(x, y)\} = \{a_1 + a_2 x + a_3 y\} = \{1\, x\, y\} \begin{Bmatrix} a_1 \\ a_2 \\ a_3 \end{Bmatrix} \quad (4\text{-}15)$$

$$\{u(x, y)\} = \{1\, x\, y\} \cdot \frac{1}{2A} \begin{bmatrix} \alpha_i & \alpha_j & \alpha_m \\ \beta_i & \beta_j & \beta_m \\ \gamma_i & \gamma_j & \gamma_m \end{bmatrix} \begin{Bmatrix} u_i \\ u_j \\ u_m \end{Bmatrix} \quad (4\text{-}16)$$

Equation (4-16) is expanded to obtain a general expression for the x-displacement function, $u(x, y)$:

$$\{u(x, y)\} = \frac{1}{2A} \cdot (1 \, x \, y) \begin{Bmatrix} \alpha_i u_i + \alpha_j u_j + \alpha_m u_m \\ \beta_i u_i + \beta_j u_j + \beta_m u_m \\ \gamma_i u_i + \gamma_j u_j + \gamma_m u_m \end{Bmatrix} \quad (4\text{-}17)$$

$$\{u(x, y)\} = \frac{1}{2A} \cdot \{\alpha_i u_i + \alpha_j u_j + \alpha_m u_m + (\beta_i u_i + \beta_j u_j + \beta_m u_m)x$$
$$+ (\gamma_i u_i + \gamma_j u_j + \gamma_m u_m)y\} \quad (4\text{-}18)$$

Equation (4-18) is rearranged into a simple algebraic form:

$$u(x, y) = \frac{1}{2A} \{(\alpha_i + \beta_i x + \gamma_i y) \cdot u_i + (\alpha_j + \beta_j x + \gamma_j y) \cdot u_j$$
$$+ (\alpha_m + \beta_m x + \gamma_m y) \cdot u_m\} \quad (4\text{-}19)$$

The Y-displacement function $v(x, y)$ can be deducted from expression (4-19):

$$v(x, y) = \frac{1}{2A} \{(\alpha_i + \beta_i x + \gamma_i y) \cdot v_i + (\alpha_j + \beta_j x + \gamma_m y) \cdot v_j$$
$$+ (\alpha_m + \beta_m x + \gamma_m y) \cdot v_m\} \quad (4\text{-}20)$$

To express the general displacement function $\{\phi\}$ in a simpler notation, let us define the coefficients of u and v displacements by new notations.

$$\begin{aligned} N_i &= (\alpha_i + \beta_i x + \gamma_i y) \\ N_j &= (\alpha_j + \beta_j x + \gamma_j y) \\ N_m &= (\alpha_m + \beta_m x + \gamma_m y) \end{aligned} \quad (4\text{-}21)$$

Expressions (4-19) and (4-20) can be rewritten using newly defined (N) values by expressions (4-21).

$$u(x, y) = \frac{1}{2A} (N_i u_i + N_j u_j + N_m u_m)$$
$$v(x, y) = \frac{1}{2A} (N_i v_i + N_j v_j + N_m v_m) \quad (4\text{-}22)$$

Combining equations (4-2), and (4-22) and using matrix notation yields:

$$\{\phi\} = \begin{Bmatrix} u(x, y) \\ v(x, y) \end{Bmatrix} = \frac{1}{2A} \begin{Bmatrix} N_i u_i + N_j u_j + N_m u_m \\ N_i v_i + N_j v_j + N_m v_m \end{Bmatrix} \quad (4\text{-}23)$$

$$\{\phi\} = \begin{Bmatrix} u(x,y) \\ v(x,y) \end{Bmatrix} = \frac{1}{2A} \begin{bmatrix} N_i & 0 & N_j & 0 & N_m & 0 \\ 0 & N_i & 0 & N_j & 0 & N_m \end{bmatrix} \begin{Bmatrix} u_i \\ v_i \\ u_j \\ v_j \\ u_m \\ v_m \end{Bmatrix} \qquad (4\text{-}24)$$

The equation can be written in the general form

$$\{\phi\} = \frac{1}{2A} \cdot \{N\}\{\delta\} \qquad (4\text{-}25)$$

where the matrix $\{N\}$ might be considered as a position matrix. Equation (4-25) will allow expression of the general deflection $\{\phi\}$ as a function of the unknown nodal displacements.

c The element strains are expressed as function of the general displacement equation (4-25). The relationship between the strains and displacements is defined by Equation (3-8).

V.2,2,3

$$\{\epsilon\} = \begin{Bmatrix} \epsilon_x \\ \epsilon_y \\ \gamma_{xy} \end{Bmatrix} = \begin{Bmatrix} \dfrac{\partial u}{\partial x} \\ \dfrac{\partial v}{\partial y} \\ \dfrac{\partial u}{\partial y} + \dfrac{\partial v}{\partial x} \end{Bmatrix}$$

Since the values of $u(x,y)$ and $v(x,y)$ are already determined by Equation (4-22), the values of the partials in Equation (3-8) are computed.

$$\frac{\partial u}{\partial x} = \frac{\partial}{\partial x}(N_i u_i + N_j u_j + N_m u_m) \cdot \frac{1}{2A} \qquad (4\text{-}26)$$

$$\frac{\partial u}{\partial x} = \left(\frac{\partial N_i}{\partial x} \cdot u_i + \frac{\partial N_j}{\partial x} \cdot u_j + \frac{\partial N_m}{\partial x} \cdot u_m\right) \cdot \frac{1}{2A} \qquad (4\text{-}27)$$

The partials of the elements of $\{N\}$ matrix are calculated. The equations given in (4-21) are differentiated with respect to the variable x.

$$\frac{\partial N_i}{\partial x} = \frac{\partial}{\partial x}(\alpha_i + \beta_i \cdot x + \gamma_i \cdot y) = \beta_i \qquad (4\text{-}28)$$

$$\frac{\partial N_j}{\partial x} = \frac{\partial}{\partial x}(\alpha_j + \beta_j \cdot x + \gamma_j \cdot y) = \beta_j \quad (4\text{-}29)$$

$$\frac{\partial N_m}{\partial x} = \frac{\partial}{\partial x}(\alpha_m + \beta_m \cdot x + \gamma_m \cdot y) = \beta_m \quad (4\text{-}30)$$

Equation (4-27) can be rewritten using the values of Equations (4-28), (4-29), and (4-30).

$$\frac{\partial u}{\partial x} = \frac{1}{2A} \cdot (\beta_i \cdot u_i + \beta_j \cdot u_j + \beta_m \cdot u_m) \quad (4\text{-}31)$$

According to similar logic, the other two elements of Equation (3-8) are determined:

$$\frac{\partial v}{\partial y} = \frac{1}{2A} \cdot (\gamma_i \cdot v_i + \gamma_j \cdot v_j + \gamma_m \cdot v_m) \quad (4\text{-}32)$$

$$\frac{\partial u}{\partial y} + \frac{\partial v}{\partial x} = \frac{1}{2A}(\gamma_i \cdot u_i + \beta_i \cdot v_i + \gamma_j \cdot u_j + \beta_j \cdot v_j + \gamma_m \cdot u_m + \beta_m \cdot v_m) \quad (4\text{-}33)$$

Now the general strain expression is written:

$$\{\epsilon\} = \begin{Bmatrix} \epsilon_x \\ \epsilon_y \\ \gamma_{xy} \end{Bmatrix} = \frac{1}{2A} \cdot \begin{bmatrix} \beta_i & 0 & \beta_j & 0 & \beta_m & 0 \\ 0 & \gamma_i & 0 & \gamma_j & 0 & \gamma_m \\ \gamma_i & \beta_i & \gamma_j & \beta_j & \gamma_m & \beta_m \end{bmatrix} \begin{Bmatrix} u_i \\ v_i \\ u_j \\ v_j \\ u_m \\ v_m \end{Bmatrix} \quad (4\text{-}34)$$

Using partitioning concept in matrix notation, expression (4-34) is written as

$$\{\epsilon\} = \frac{1}{2A} \cdot \{B_i B_j B_m\} \begin{Bmatrix} \delta_i \\ \delta_j \\ \delta_m \end{Bmatrix} \quad (4\text{-}35)$$

which can be simply rewritten in a concise and general matrix form:

$$\{\epsilon\} = \{B\}\{\delta\} \quad (4\text{-}36)$$

where $\{\epsilon\}$ is the general strain matrix for the element, $\{\delta\}$ the nodal displacements matrix, and $\{B\}$ the displacement-strain matrix whose elements are independent of the coordinates X, Y of the system as shown in Equation (4-10). Therefore the strains given by Equation (4-36) have constant values which are consistent with the definition of the constant-strain triangular element.

Besides the strains $\{\epsilon\}$ which are associated with the displacement functions, there are others such as the ones due to temperature variations, $\{\epsilon_T\}$ and initially present ones, $\{\epsilon_0\}$. In matrix notation they will be written as

$$\{\epsilon_0\} = \begin{Bmatrix} \epsilon_{x0} \\ \epsilon_{y0} \\ \gamma_{xy0} \end{Bmatrix} \quad \text{and} \quad \{\epsilon_T\} = \begin{Bmatrix} \epsilon_{xT} \\ \epsilon_{yT} \\ \gamma_{xyT} \end{Bmatrix} \tag{4-37}$$

To determine the final strain values, $\{\epsilon_0\}$ and $\{\epsilon_T\}$ of the expression (4-37) need to be algebraically added to the value of Equation (4-36).

$$\{\epsilon\}_{\text{final}} = \{\epsilon\} + \{\epsilon_0\} + \{\epsilon_T\} \tag{4-38}$$

d The general stress expression is related to the strain relations of the element by a transformation matrix $\{D\}$, known as the *elasticity matrix* [2]. Equation (3-12), which relates the strains and stresses, can be rewritten as

$$\{\sigma\} = \begin{Bmatrix} \sigma_x \\ \sigma_y \\ \tau_{xy} \end{Bmatrix} = [D] \begin{Bmatrix} \epsilon_x \\ \epsilon_y \\ \gamma_{xy} \end{Bmatrix} \tag{4-39}$$

In the analysis of plane stress problems dealing with isotropic material, the stress-strain relationships are given by

$$\epsilon_x = \frac{\sigma_x}{E} - \nu \frac{\sigma_y}{E}$$

$$\epsilon_y = \frac{\sigma_y}{E} - \nu \frac{\sigma_x}{E} \tag{4-40}$$

$$\gamma_{xy} = 2(1 + \nu) \frac{\tau_{xy}}{E}$$

Equation (4-40) in matrix notation will have the following form:

$$\begin{Bmatrix} \epsilon_x \\ \epsilon_y \\ \gamma_{xy} \end{Bmatrix} = \frac{1}{E} \begin{bmatrix} 1 & -\nu & 0 \\ -\nu & 1 & 0 \\ 0 & 0 & 2(1+\nu) \end{bmatrix} \begin{Bmatrix} \sigma_x \\ \sigma_y \\ \tau_{xy} \end{Bmatrix} \tag{4-41}$$

The matrix equation (4-41) is rearranged to solve for the stresses $\{\sigma\}$:

$$\begin{Bmatrix} \sigma_x \\ \sigma_y \\ \tau_{xy} \end{Bmatrix} = E \begin{bmatrix} 1 & -\nu & 0 \\ -\nu & 1 & 0 \\ 0 & 0 & 2(1+\nu) \end{bmatrix}^{-1} \begin{Bmatrix} \epsilon_x \\ \epsilon_y \\ \gamma_{xy} \end{Bmatrix} \quad (4\text{-}42)$$

The inverse operation is performed to rewrite Equation (4-42):

$$\begin{Bmatrix} \sigma_x \\ \sigma_y \\ \tau_{xy} \end{Bmatrix} = \frac{E}{1-\nu^2} \begin{bmatrix} 1 & \nu & 0 \\ \nu & 1 & 0 \\ 0 & 0 & \frac{1-\nu}{2} \end{bmatrix} \begin{Bmatrix} \epsilon_x \\ \epsilon_y \\ \gamma_{xy} \end{Bmatrix} \quad (4\text{-}43)$$

Equation (4-43) is identical to Equation (3-12), which defines the elasticity stress matrix $\{D\}$:

$$\{D\} = \frac{E}{1-\nu^2} \begin{bmatrix} 1 & \nu & 0 \\ \nu & 1 & 0 \\ 0 & 0 & \frac{1-\nu}{2} \end{bmatrix} \quad (4\text{-}44)$$

in which ν is the Poisson ratio and E is Young's modulus.

In the analysis of plane strain problems dealing with isotropic material, the stress-strain relationships are given by the following equations.

$$\epsilon_x = \frac{\sigma_x}{E} - \nu \frac{\sigma_y}{E} - \nu \frac{\sigma_z}{E} \quad (4\text{-}45)$$

$$\epsilon_y = -\nu \frac{\sigma_x}{E} + \frac{\sigma_y}{E} - \nu \frac{\sigma_z}{E} \quad (4\text{-}46)$$

$$\epsilon_z = -\frac{\nu \sigma_x}{E} - \frac{\nu \sigma_y}{E} + \frac{\sigma_z}{E} = 0 \quad (4\text{-}47)$$

$$\gamma_{xy} = 2(1+\nu) \frac{\tau_{xy}}{E} \quad (4\text{-}48)$$

In this case, since ϵ_z is zero, the normal stress component σ_z exists. The procedure of solution will determine σ_z from equation (4-47).

$$\sigma_z = \nu(\sigma_x + \sigma_y) \quad (4\text{-}49)$$

The new forms of Equations (4-45) and (4-46) will be obtained by substituting Equation (4-49) in them:

$$\epsilon_x = \frac{\sigma_x}{E} - \nu \frac{\sigma_y}{E} - \frac{\nu}{E} \{\nu(\sigma_x + \sigma_y)\}$$

$$\epsilon_y = -\frac{\nu\sigma_x}{E} + \frac{\sigma_y}{E} - \frac{\nu}{E}\{\nu(\sigma_x + \sigma_y)\} \qquad (4\text{-}50)$$

$$\gamma_{xy} = 2(1 + \nu)\frac{\tau_{xy}}{E}$$

Expressions (4-50) can be rearranged by collecting similar terms and written using matrix notation.

$$\begin{Bmatrix} \epsilon_x \\ \epsilon_y \\ \gamma_{xy} \end{Bmatrix} = \begin{bmatrix} \dfrac{1-\nu^2}{E} & \dfrac{-\nu-\nu^2}{E} & 0 \\ \dfrac{-\nu-\nu^2}{E} & \dfrac{1-\nu^2}{E} & 0 \\ 0 & 0 & \dfrac{2(1+\nu)}{E} \end{bmatrix} \begin{Bmatrix} \sigma_x \\ \sigma_y \\ \tau_{xy} \end{Bmatrix} \qquad (4\text{-}51)$$

Equation (4-51) needs to be written for direct solution of stresses, $\{\sigma\}$:

$$\begin{Bmatrix} \sigma_x \\ \sigma_y \\ \tau_{xy} \end{Bmatrix} = \begin{bmatrix} \dfrac{1-\nu^2}{E} & \dfrac{-\nu-\nu^2}{E} & 0 \\ \dfrac{-\nu-\nu^2}{E} & \dfrac{1-\nu^2}{E} & 0 \\ 0 & 0 & \dfrac{2(1+\nu)}{E} \end{bmatrix}^{-1} \begin{Bmatrix} \epsilon_x \\ \epsilon_y \\ \gamma_{xy} \end{Bmatrix} \qquad (4\text{-}52)$$

The inversion of the above matrix may be performed by the adjoint inversion method, Equation (1-9).

$$\text{Adj}\begin{bmatrix} \dfrac{1-\nu^2}{E} & \dfrac{-\nu-\nu^2}{E} & 0 \\ \dfrac{-\nu-\nu^2}{E} & \dfrac{1-\nu^2}{E} & 0 \\ 0 & 0 & \dfrac{2(1+\nu)}{E} \end{bmatrix}$$

$$= \begin{bmatrix} \dfrac{1-\nu^2}{E}\cdot\dfrac{2(1+\nu)}{E} & \dfrac{\nu+\nu^2}{E}\cdot\dfrac{2(1+\nu)}{E} & 0 \\ \dfrac{\nu+\nu^2}{E}\cdot\dfrac{2(1+\nu)}{E} & \dfrac{1-\nu^2}{E}\cdot\dfrac{2(1+\nu)}{E} & 0 \\ 0 & 0 & \left(\dfrac{1-\nu^2}{E}\right)^2 - \left(\dfrac{\nu+\nu^2}{E}\right)^2 \end{bmatrix} \qquad (4\text{-}53)$$

PLANE STRESS AND PLANE STRAIN ANALYSIS

$$\text{Determinant} = \frac{2(1+\nu)}{E}\left\{\left(\frac{1-\nu^2}{E}\right)^2 - \left(\frac{\nu+\nu^2}{E}\right)^2\right\} \quad (4\text{-}54)$$

Then the elasticity matrix for the plane strain problem, using isotropic material, becomes

$$\{D\} = \frac{E(1-\nu)}{(1+\nu)(1-2\nu)} \begin{bmatrix} 1 & \frac{\nu}{1-\nu} & 0 \\ \frac{\nu}{1-\nu} & 1 & 0 \\ 0 & 0 & \frac{(1-2\nu)}{2(1-\nu)} \end{bmatrix} \quad (4\text{-}55)$$

Therefore the $\{D\}$ matrix of the general equation (3-11) has two forms shown by expressions (4-44) and (4-55). They should be associated with the corresponding analysis problems.

e The displacement finite element analysis uses the nodal displacements as the unknowns of the problem. They are determined by solving the system of linear equations generated by satisfying the equilibrium conditions at the nodes. The stationary total potential energy principle is the foundation of this analysis. The total potential energy of the system is given by Equation (3-1).

$$(V) = (U) - (U_F) - (U_R) - (U_S)$$

The generalized expression for each element of Equation (3-1) are given below.

a Internal strain energy expression, (U).

$$(U) = \tfrac{1}{2}\int_V \{\epsilon\}^T\{\sigma\}\,dV \quad (4\text{-}56)$$

If the equation (4-39) is substituted in the expression (4-56), one obtains

$$(U) = \tfrac{1}{2}\int_V \{\epsilon\}^T[D]\{\epsilon\}\,dV \quad (4\text{-}57)$$

b Potential energy due to body forces, (U_F).

$$(U_F) = \int_V \rho\{\phi\}^T\{F\}\,dV \quad (4\text{-}58)$$

where $\{F\}$ is the body forces of the system, ρ the mass density of the elastic body, and $\{\phi\}$ the general displacement function.

c Potential energy due to the external nodal forces (U_R).

$$(U_R) = \{\delta\}^T\{R\} \tag{4-59}$$

in which $\{\delta\}$ is the nodal displacements and $\{R\}$ is the external nodal loads.

d Potential energy due to the surface traction (U_S).

$$(U_S) = \int_{area} \{\phi\}^T\{S\}\, dA \tag{4-60}$$

where $\{\phi\}$ is the general displacement function and $\{S\}$ is the surface tractions. Equation (4-60) can be rewritten in the following manner, using Equation (4-25).

$$(U_S) = \int_{area} \frac{1}{2A} \cdot \{\delta\}^T\{N\}^T\{S\}\, dA \tag{4-61}$$

Expression (3-1) can be expanded by substituting in it the values given in Equations (4-57), (4-58), (4-59), and (4-60).

$$(V) = \tfrac{1}{2}\int_V \{\epsilon\}^T[D]\{\epsilon\}\, dV - \int_V \rho\cdot\{\phi\}^T\{F\}\, dV - \{\delta\}^T\{R\}$$

$$- \int_{area} \frac{1}{2A} \cdot \{\delta\}^T\{N\}^T\{S\}\, dA \tag{4-62}$$

Substituting Equations (4-36) and (4-25) in expression (4-62), one can write

$$(V) = \tfrac{1}{2}\int_V \{\delta\}^T\{B\}^T[D]\{B\}\{\delta\}\, dV - \int_V \frac{1}{2A}\cdot\rho\{\delta\}^T\{N\}^T\{F\}\, dV$$

$$- \{\delta\}^T\{R\} - \int_{area} \frac{1}{2A}\cdot\{\delta\}^T\{N\}^T\{S\}\, dA \tag{4-63}$$

If the structural system which is being analyzed has n elements, then expression (4-63) needs to be summed over n elements.

$$(V) = \sum_{1}^{n}\left[\tfrac{1}{2}\int_V \{\delta\}^T\{B\}^T[D]\{B\}\{\delta\}\, dV - \int_V \frac{1}{2A}\cdot\rho\{\delta\}^T\{N\}^T\{F\}\, dV\right.$$

$$\left. - \{\delta\}^T\{R\} - \int_{area} \frac{1}{2A}\cdot\{\delta\}^T\{N\}^T\{S\}\, dA\right] \tag{4-64}$$

The nodal displacements $\{\delta\}$ are independent of the general coordinates x and y; therefore they can be taken out of the integrals of the equation (4-64):

$$(V) = \sum_{1}^{n} \left[\tfrac{1}{2}\{\delta\}^T \int_V \{B\}^T[D]\{B\}\, dV \cdot \{\delta\} - \{\delta\}^T \int_V \frac{\rho}{2A} \{N\}^T\{F\}\, dV \right.$$
$$\left. - \{\delta\}^T\{R\} - \{\delta\}^T \int_{\text{area}} \frac{1}{2A} \cdot \{N\}^T\{S\}\, dA \right] \quad (4\text{-}65)$$

It is noted that the last three terms of expression (4-65) represent the total loading system $\{Q\}$ of the analysis.

$$\{Q\} = \sum_{1}^{n} \int_V \frac{\rho}{2A} \{N\}^T\{F\}\, dV + \{R\} + \int_{\text{area}} \frac{1}{2A} \{N\}^T\{S\}\, dA \quad (4\text{-}66)$$

Then Equation (4-65) is rewritten:

$$(V) = \tfrac{1}{2}\{\delta\}^T \int_V \{B\}^T[D]\{B\}\, dV \cdot \{\delta\} - \{\delta\}^T\{Q\} \quad (4\text{-}67)$$

The integral $\left\{ \int_V \{B\}^T[D]\{B\}\, dV \right\}$ defines the stiffness matrix $[k]$ of an element [3].

$$[k] = \int_V \{B\}^T[D]\{B\}\, dV \quad (4\text{-}68)$$

The superposition of all the element stiffness matrices will yield the total structural stiffness matrix $[K]$ of the system.

$$[K] = \sum_{1}^{n} [k] \quad (4\text{-}69)$$

Now Equation (4-67) can be simplified by substituting expression (4-69) in it.

$$(V) = \tfrac{1}{2}\{\delta\}^T[K]\{\delta\} - \{\delta\}^T\{Q\} \quad (4\text{-}70)$$

According to the principle of stationary potential energy, the equilibrium conditions are satisfied when this energy is at a minimum. This concept is applied to the problem by letting the partial derivatives of the total potential energy with respect to the unknown nodal displacements to vanish as shown in Equation (3-3). The defined mathematical operations on expression (4-70) will generate the necessary system of linear equations which express the nodal equilibrium conditions.

$$[K]\{\delta\} = \{Q\} \quad (4\text{-}71)$$

which is rearranged to solve for the unknown nodal displacements $\{\delta\}$:

$$\{\delta\} = [K]^{-1}\{Q\} \tag{4-72}$$

Equation (4-72) was previously expressed by Equation (3-15).

Once the unknown nodal displacements are computed, then the strains are calculated by Equation (4-36), which is

$$\{\epsilon\} = \{B\}\{\delta\}$$

and the stresses are computed by Equation (4-39), which is

$$\{\sigma\} = [D]\{\epsilon\}$$

This will terminate the logical development of the finite element method using a constant-strain triangular element to solve a plane stress problem.

Stiffness Matrix

Equation (4-69) defined the total stiffness matrix of the structural system as

$$[K] = \sum_{1}^{n} [k]$$

where $[k]$ is the stiffness matrix of one element defined by expression (4-68) as

$$[k] = \int_{V} \{B\}^{T}[D]\{B\} \, dV$$

Since the matrix quantities $\{B\}$ and $[D]$ are independent of the coordinates X and Y, expression (4-68) can be written as

$$[k] = \{B\}^{T}[D]\{B\} \cdot t \cdot \frac{dx \cdot dy}{2} \tag{4-73}$$

where (t) is the element thickness; $(dx \cdot dy)$ is twice the area of the element. The expansion of Equation (4-73) will yield, for one triangular element, the stiffness matrix in the following matrix form:

$$[k] = \begin{bmatrix} K_{ii} & K_{ij} & K_{im} \\ K_{ji} & K_{jj} & K_{jm} \\ K_{mi} & K_{mj} & K_{mm} \end{bmatrix} \tag{4-74}$$

where each element of the matrix $[k]$ is a 2×2 matrix given by the general expression,

$$[K_{ij}] = \{B_{i}\}^{T}[D]\{B_{j}\} \cdot t \cdot \text{area} \tag{4-75}$$

The expressions (4-73) can be rewritten in an expanded form as

PLANE STRESS AND PLANE STRAIN ANALYSIS

$$[k] = \{B_i B_j B_m\}^T [D] \{B_i B_j B_m\} \cdot t \cdot \text{area} \quad (4\text{-}76)$$

If the matrix multiplications are performed, the result will be a 3 × 3 matrix for the element stiffness matrix $[k]$:

$$[k] = \begin{bmatrix} \{B_i\}^T[D]\{B_i\} & \{B_i\}^T[D]\{B_j\} & \{B_i\}^T[D]\{B_m\} \\ \{B_j\}^T[D]\{B_i\} & \{B_j\}^T[D]\{B_j\} & \{B_j\}^T[D]\{B_m\} \\ \{B_m\}^T[D]\{B_i\} & \{B_m\}^T[D]\{B_j\} & \{B_m\}^T[D]\{B_m\} \end{bmatrix} \cdot t \cdot \text{area} \quad (4\text{-}77)$$

To develop the total stiffness matrix of a structural system, the element stiffness matrices are properly superimposed. To illustrate this procedure, a system with two triangular elements is considered, Fig. 4-2. The element

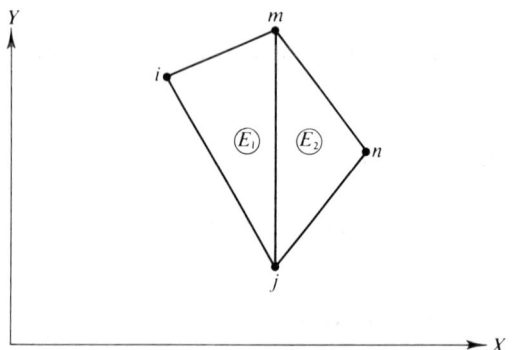

Figure 4-2

stiffness matrix for the element E_1, according to Equation (4-74), will be

$$[k_{E_1}] = \begin{bmatrix} K_{ii} & K_{ij} & K_{im} \\ K_{ji} & K_{jj} & K_{jm} \\ K_{mi} & K_{mj} & K_{mm} \end{bmatrix} \quad (4\text{-}78)$$

The element stiffness matrix for the element E_2, according to Equation (4-74), will be

$$[k_{E_2}] = \begin{bmatrix} K_{jj} & K_{jn} & K_{jm} \\ K_{nj} & K_{nn} & K_{nm} \\ K_{mj} & K_{mn} & K_{mm} \end{bmatrix} \quad (4\text{-}79)$$

Expressions (4-78) and (4-79) are superimposed to obtain the total stiffness matrix $[K]$ of the system.

$$[K] = \begin{bmatrix} (K_{ii})_1 & (K_{ij})_1 & 0 & (K_{im})_1 \\ (K_{ji})_1 & (K_{jj})_1 + (K_{jj})_2 & (K_{jn})_2 & (K_{jm})_1 + (K_{jm})_2 \\ 0 & (K_{nj})_2 & (K_{nn})_2 & (K_{nm})_2 \\ (K_{mi})_1 & (K_{mj})_1 + (K_{mj})_2 & (K_{mn})_2 & (K_{mm})_1 + (K_{mm})_2 \end{bmatrix} \quad (4\text{-}80)$$

The stiffness matrix $[K]$ is a square, symmetric, and positive-definite matrix. The solution of the linear equations related to the finite element analysis takes up to one-half of the total computer usage time. The improvement and simplification of the stiffness matrix $[K]$ will increase the efficiency of the computer usage and decrease the computer-run time. This can be accomplished by the use of the characteristics of the stiffness matrix.

It is also a fact that the matrix $[K]$ is often a well-banded one. Most of its elements are located at the neighborhood of the diagonal. Banding is a desirable quality of a matrix, since only the elements in the band will be stored in the computer and they are the ones to be used for the solution operations. It has been shown that the solution time of matrix equation systems varies approximately as the square of the bandwidth of the matrix quantities. Hence there is a need to find means to decrease the bandwidths.

The bandwidth of a stiffness matrix is associated with the type of structure being analyzed and also with the scheme used in numbering the nodes. Since the given structural system cannot be altered, the only procedure to decrease the bandwidth is the improvement of the numbering of the nodes of the elements. This is accomplished by minimizing the difference between two adjacent nodal point numbers. To illustrate this concept, let us consider a structural system with two elements and number their nodes with two different configurations.

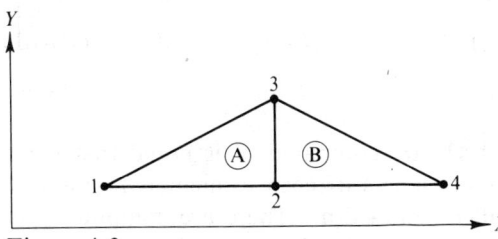

Figure 4-3

Case 1. The difference between adjacent nodes is minimized Fig. 4-3. The total stiffness matrix $[K]_1$ will be written according to scheme shown by Equation (4-80).

$$[K]_1 = \begin{matrix} & 1 & 2 & 3 & 4 \\ 1 \\ 2 \\ 3 \\ 4 \end{matrix} \begin{bmatrix} (K_{11})_A & (K_{12})_A & (K_{13})_A & 0 \\ (K_{21})_A & (K_{22})_A + (K_{22})_B & (K_{23})_A + (K_{23})_B & (K_{24})_B \\ (K_{31})_A & (K_{32})_A + (K_{32})_B & (K_{33})_A + (K_{33})_B & (K_{34})_B \\ 0 & (K_{42})_B & (K_{43})_B & (K_{44})_B \end{bmatrix} \quad (4\text{-}81)$$

Case 2. The difference between adjacent nodes is not minimized,

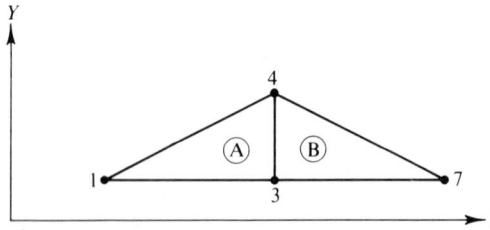

Figure 4-4

Fig. 4-4. The total stiffness matrix $[K]_2$ will be written according to the scheme shown by Equation (4-80).

$$[K]_2 = \begin{matrix} & 1 & 2 & 3 & 4 & 5 & 6 & 7 \\ 1 \\ 2 \\ 3 \\ 4 \\ 5 \\ 6 \\ 7 \end{matrix} \begin{bmatrix} (K_{11})_A & 0 & (K_{13})_A & (K_{14})_A & 0 & 0 & 0 \\ 0 & 0 & 0 & 0 & 0 & 0 & 0 \\ (K_{31})_A & 0 & (K_{33})_A + (K_{33})_B & (K_{34})_A + (K_{34})_B & 0 & 0 & (K_{37})_B \\ (K_{41})_A & 0 & (K_{43})_A + (K_{43})_B & (K_{44})_A + (K_{44})_B & 0 & 0 & (K_{47})_B \\ 0 & 0 & 0 & 0 & 0 & 0 & 0 \\ 0 & 0 & 0 & 0 & 0 & 0 & 0 \\ 0 & 0 & (K_{73})_B & (K_{74})_B & 0 & 0 & (K_{77})_B \end{bmatrix}$$

$$(4\text{-}82)$$

When expressions (4-81) and (4-82) are compared, it is noted that when the difference between the nodal numbers of each element is minimized, the stiffness matrix has a smaller bandwidth. There are computer programs which renumber the nodes to generate the most efficient structural stiffness matrix of the system [4]. This theoretical development of the finite element method is numerically explained by the formulation of problems in Chapter 5. It might be helpful to study the material in Chapters 4 and 5 in a parallel and correlated manner.

PROBLEMS

4-1 Using the concepts of the theory of elasticity, explain the conditions and requirements in order to classify a problem as plane stress or plane strain case. Give rigorous mathematical explanations.

4-2 Make a list of all matrix quantities which are relevant to the stiffness finite element analysis of a plane stress analysis problem. Define them properly and explain their interrelations.

4-3 Perform the longhand stiffness finite element method analysis of the given plane truss. Generate all the necessary matrices with explanations and reference to the proper matrix expressions presented in the text. This is a theoretical analysis problem to illustrate the complete logic of the finite element method. Assume AE/L is constant. The geometry of the truss is given by Fig. 4-5.

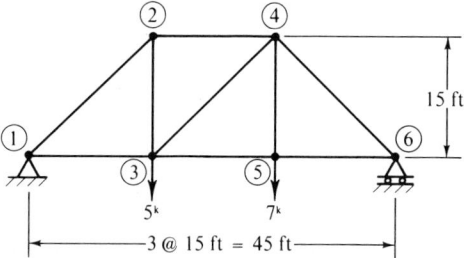

Figure 4-5

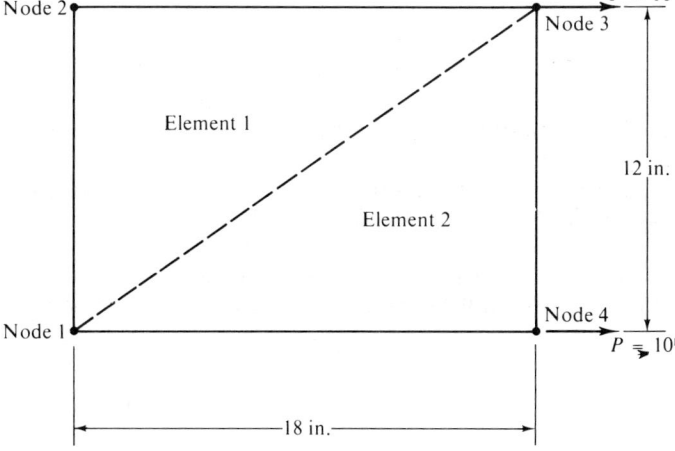

Figure 4-6

4-4 Analyze the thin plate shown in Fig. 4-6 by the finite element method. Perform all the operations in longhand to illustrate the logic of the method. This is a simple plane stress problem, since the load creates a pure tension stress condition in the plate. For simplicity, use only two constant-strain triangles to generate the rough mesh. The plate is fixed at one end and loaded at the opposite end as shown in Fig. 4-6. Young's modulus is 30,000 ksi and Poisson's ratio is 0.3. Plate thickness is 2 in.

4-5 Perform the longhand stiffness finite element method analysis of the given plane truss. The geometry and loading of the truss is shown by

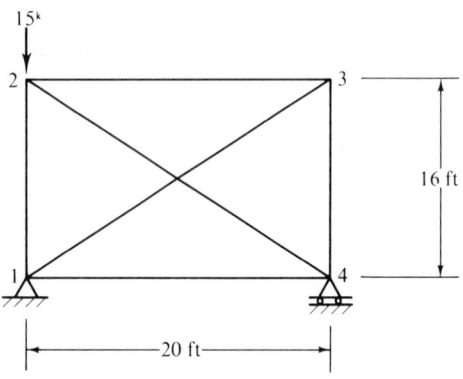

Figure 4-7

Fig. 4-7. The truss members are made of A-36 steel and they all have cross-sectional areas of 3 sq in.

REFERENCES

1. J. J. TURNER, R. W. CLOUGH, H. C. MARTIN, and L. J. TOPP, "Stiffness and Deflection Analysis of Complex Structures," *Journal of Aeronautical Sciences* **23**, 805–832 (1956).
2. S. TIMOSHENKO and J. H. GOODIER, *Theory of Elasticity*. New York: McGraw-Hill, 1970.
3. R. W. CLOUGH, and Y. RASHID, "Finite Element Analysis of Axi-Symmetric Solids," *Journal of the Engineering Mechanics Division, Proceedings of the ASCE* 76, (February 1965).
4. F. A. AKYÜZ and S. UTKU, "An Automatic Node Relabelling Scheme for Bandwidth Minimization of Stiffness Matrices," *AIAA Journal* **6**, 4, 728–730 (April 1968).

Longhand Solution of Plane Stress Problems

INTRODUCTION

The theoretical development of the finite element method becomes more meaningful and gains a new dimension in practical application when it is used for the numerical solution of a simple problem. To illustrate the application procedure, three types of problems are considered:

a   A plane truss
b   A plate in pure tension
c   A simply supported beam

The reader should remember that these longhand solutions are presented to give a complete and solid understanding of the finite element method. The actual formulations will definitely need the help of the high-speed electronic digital computers to perform the necessary matrix operations. The general logic used for the formulation of the three problems are similar and will have the following major steps.

1. The idealization of the structure by the selection of element type and element size. (This is known as the *mesh generation* for the system.)
2. Definition of the geometric and elastic properties of the system.
3. Definition of the boundary conditions of the system.
4. Assumption of general displacement functions to describe the displacement behavior of the selected elements.
5. Generation of the element stiffness matrices, using the principle of stationary potential energy.
6. The superposition of the element stiffness matrices to develop the total structural stiffness matrix.
7. Solution of the system of linear equations generated by the satisfaction of the equilibrium conditions at all nodal points. This will yield the values of the unknown nodal displacements.
8. Calculation of element strains, using the computed nodal displacement values.
9. Calculation of element stresses, using either the known nodal point displacements or element strains.
10. Calculation of the principal stresses, using the known values of element stresses.

Example 5-1

Perform the structural analysis of the plane truss shown in Fig. 5-1 using the finite element method. The Young's modulus E and L/A are constant.

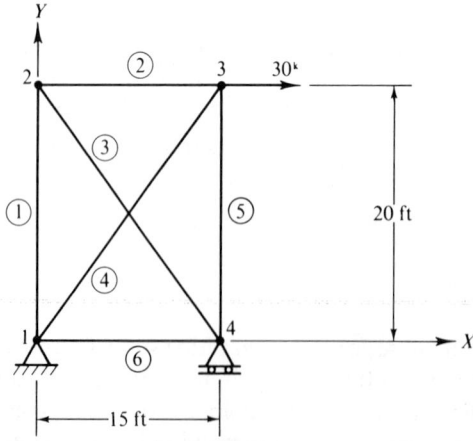

Figure 5-1

General Discussion

The plane truss has only linear elements with two nodal points. The finite element analysis of such problems will be simpler than for two dimensional problems.

It will be proper to present a general discussion on the characteristics of linear members with various types of possible loading conditions. An inclined truss member is shown in Fig. 5-2.

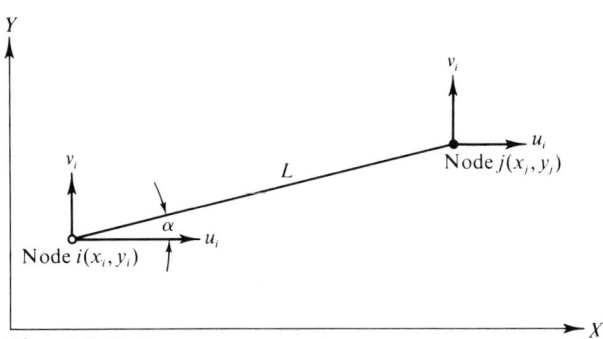

Figure 5-2

The length of the member is given by the following relation.

$$L = \sqrt{(x_j - x_i)^2 + (y_j - y_i)^2} \qquad (5\text{-}1)$$

The angle of inclination of the member is given by the angle α.

$$\alpha = \tan^{-1}\left\{\frac{y_j - y_i}{x_j - x_i}\right\} \qquad (5\text{-}2)$$

Each node of the linear element will have two degrees of freedom. The general nodal displacement matrix for an element will be

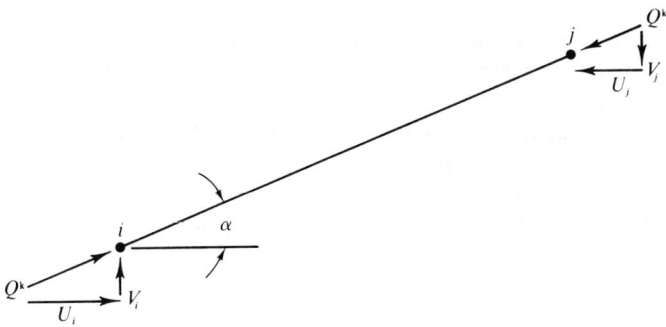

Figure 5-3

$$\{\delta\} = \begin{Bmatrix} \delta_i \\ \delta_j \end{Bmatrix} = \begin{Bmatrix} u_i \\ v_i \\ u_j \\ v_j \end{Bmatrix} \quad (5\text{-}3)$$

where u_i and u_j are the x displacements and v_i and v_j are the y displacements of the nodes i and j.

The nodal forces $\{P\}$ for a linear element can be written as

$$\{P\} = \begin{Bmatrix} P_i \\ P_j \end{Bmatrix} = \begin{Bmatrix} U_i \\ V_i \\ U_j \\ V_j \end{Bmatrix} \text{ kips} \quad (5\text{-}4)$$

The nodal displacements given in expression (5-3) will cause member deformations in the form of elongations. When these member elongations are multiplied by (AE/L), the axial member forces are obtained. The components of the axial member force in the X and Y directions are shown in Fig. 5-3.

$$\{P\} = \begin{Bmatrix} U_i \\ V_i \\ U_j \\ V_j \end{Bmatrix} = \begin{Bmatrix} \cos\alpha \\ \sin\alpha \\ -\cos\alpha \\ -\sin\alpha \end{Bmatrix} \{Q\} \quad (5\text{-}5)$$

where $\{Q\}$ is the axial force on the member. The nodal displacements $\{\delta\}$ will induce member deformations which are illustrated geometrically by the Fig. 5-4.

Each node of the linear element has two degrees of freedom; therefore each linear element will have four possible degrees of freedom.

Using the geometric relations of Fig. 5-4, axial elongations are related to the nodal displacements. The general elongation will be given by the expression

Axial elongation of a member $(ij) = L_{ij} = (u_j - u_i)\cos\alpha + (v_j - v_i)\sin\alpha \quad (5\text{-}6)$

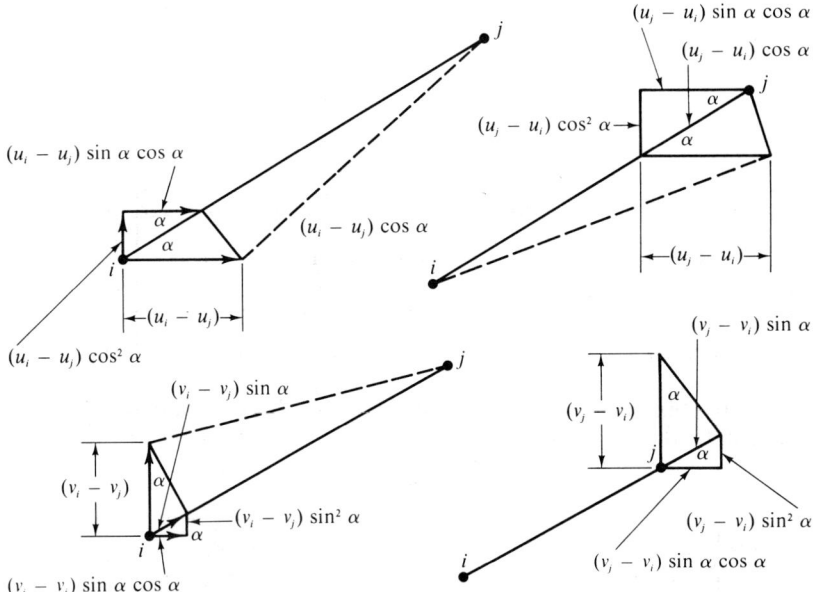

Figure 5-4

Equation (5-6) is written in matrix notation as

$$\{L_{ij}\} = \{\cos \alpha \quad \sin \alpha \quad -\cos \alpha \quad -\sin \alpha\} \begin{Bmatrix} u_i \\ v_i \\ u_j \\ v_j \end{Bmatrix} \quad (5\text{-}7)$$

To obtain the horizontal and vertical components of the axial elongation $\{L_{ij}\}$, expression (5-7) is multiplied by the coefficient matrix given in expression (5-5):

$$\begin{Bmatrix} (L_{ij})_{ix} \\ (L_{ij})_{iy} \\ (L_{ij})_{jx} \\ (L_{ij})_{jy} \end{Bmatrix} = \begin{Bmatrix} \cos \alpha \\ \sin \alpha \\ -\cos \alpha \\ -\sin \alpha \end{Bmatrix} \{\cos \alpha \quad \sin \alpha \quad -\cos \alpha \quad -\sin \alpha\} \begin{Bmatrix} u_i \\ v_i \\ u_j \\ v_j \end{Bmatrix} \quad (5\text{-}8)$$

The matrix expression (5-8) is expanded into a new form given by matrix (5-9):

$$\begin{Bmatrix} (L_{ij})_{ix} \\ (L_{ij})_{iy} \\ (L_{ij})_{jx} \\ (L_{ij})_{jy} \end{Bmatrix} = \begin{bmatrix} \cos^2 \alpha & \sin \alpha \cos \alpha & -\cos^2 \alpha & -\sin \alpha \cos \alpha \\ \sin \alpha \cos \alpha & \sin^2 \alpha & -\sin \alpha \cos \alpha & -\sin^2 \alpha \\ -\cos^2 \alpha & -\sin \alpha \cos \alpha & \cos^2 \alpha & \sin \alpha \cos \alpha \\ -\sin \alpha \cos \alpha & -\sin^2 \alpha & \sin \alpha \cos \alpha & \sin^2 \alpha \end{bmatrix} \begin{Bmatrix} u_i \\ v_i \\ u_j \\ v_j \end{Bmatrix} \quad (5\text{-}9)$$

If the matrix expression (5-9) is multiplied by the constant (EA/L), the results will be the X and Y components of the member forces at the nodes:

$$\{P\} = \begin{Bmatrix} U_i \\ V_i \\ U_j \\ V_j \end{Bmatrix}$$

$$= \frac{EA}{L} \begin{bmatrix} \cos^2 \alpha & \sin \alpha \cos \alpha & -\cos^2 \alpha & -\sin \alpha \cos \alpha \\ \sin \alpha \cos \alpha & \sin^2 \alpha & -\sin \alpha \cos \alpha & -\sin^2 \alpha \\ -\cos^2 \alpha & -\sin \alpha \cos \alpha & \cos^2 \alpha & \sin \alpha \cos \alpha \\ -\sin \alpha \cos \alpha & -\sin^2 \alpha & \sin \alpha \cos \alpha & \sin^2 \alpha \end{bmatrix} \begin{Bmatrix} u_i \\ v_i \\ u_j \\ v_j \end{Bmatrix}$$

$$(5\text{-}10)$$

Solution

1. Since the numerical calculations will use trigonometric values, it will be proper to determine them. They are given by Table 5-1.

TABLE 5-1

Member	$\cos^2 \alpha$	$\sin^2 \alpha$	$\sin \alpha \cos \alpha$
1	0	1	0
2	1	0	0
3	0.36	0.64	−0.48
4	0.36	0.64	0.48
5	0	1	0
6	1	0	0

INTRODUCTION 105

2. The relations between member-end forces and member deformations are defined by the general expressions (2-34) and (2-35). The deformation-force matrix quantities are generated for each linear element. Later, they will be superposed to develop the total structural stiffness matrix.

(a) Member 1

$$\begin{Bmatrix} U_1 \\ V_1 \\ U_2 \\ V_2 \end{Bmatrix} = \frac{EA}{L} \begin{bmatrix} K_{11} & K_{12} \\ K_{21} & K_{22} \end{bmatrix} \begin{Bmatrix} u_1 \\ v_1 \\ u_2 \\ v_2 \end{Bmatrix} \quad (5\text{-}11)$$

Each element of the matrix $[k]$ is a 2×2 matrix, hence, the actual dimension of $[k]$ is 4×4. The numerical value of the expression of (5-11) will be written as shown by expression (5-12).

$$\begin{Bmatrix} U_1 \\ V_1 \\ U_2 \\ V_2 \end{Bmatrix} = \frac{EA}{L} \begin{bmatrix} 0 & 0 & 0 & 0 \\ 0 & 1 & 0 & -1 \\ 0 & 0 & 0 & 0 \\ 0 & -1 & 0 & 1 \end{bmatrix} \begin{Bmatrix} u_1 \\ v_1 \\ u_2 \\ v_2 \end{Bmatrix} \quad (5\text{-}12)$$

(b) Member 2

$$\begin{Bmatrix} U_2 \\ V_2 \\ U_3 \\ V_3 \end{Bmatrix} = \frac{EA}{L} \begin{bmatrix} K_{22} & K_{23} \\ K_{32} & K_{33} \end{bmatrix} \begin{Bmatrix} u_2 \\ v_2 \\ u_3 \\ v_3 \end{Bmatrix} \quad (5\text{-}13)$$

The numerical value of the expression (5-13) is

$$\begin{Bmatrix} U_2 \\ V_2 \\ U_3 \\ V_3 \end{Bmatrix} = \frac{EA}{L} \begin{bmatrix} 1 & 0 & -1 & 0 \\ 0 & 0 & 0 & 0 \\ -1 & 0 & 1 & 0 \\ 0 & 0 & 0 & 0 \end{bmatrix} \begin{Bmatrix} u_2 \\ v_2 \\ u_3 \\ v_3 \end{Bmatrix} \quad (5\text{-}14)$$

(c) Member 3
For the following members only the numerical matrix quantities are given.

$$\begin{Bmatrix} U_4 \\ V_4 \\ U_2 \\ V_2 \end{Bmatrix} = \frac{EA}{L} \begin{bmatrix} 0.36 & -0.48 & -0.36 & 0.48 \\ -0.48 & 0.64 & 0.48 & -0.64 \\ -0.36 & 0.48 & 0.36 & -0.48 \\ 0.48 & -0.64 & -0.48 & 0.64 \end{bmatrix} \begin{Bmatrix} u_4 \\ v_4 \\ u_2 \\ v_2 \end{Bmatrix} \quad (5\text{-}15)$$

(d) Member 4

$$\begin{Bmatrix} U_1 \\ V_1 \\ U_3 \\ V_3 \end{Bmatrix} = \frac{EA}{L} \begin{bmatrix} 0.36 & 0.48 & -0.36 & -0.48 \\ 0.48 & 0.64 & -0.48 & -0.64 \\ -0.36 & -0.48 & 0.36 & 0.48 \\ -0.48 & -0.64 & 0.48 & 0.64 \end{bmatrix} \begin{Bmatrix} u_1 \\ v_1 \\ u_3 \\ v_3 \end{Bmatrix} \quad (5\text{-}16)$$

(e) Member 5

$$\begin{Bmatrix} U_4 \\ V_4 \\ U_3 \\ V_3 \end{Bmatrix} = \frac{EA}{L} \begin{bmatrix} 0 & 0 & 0 & 0 \\ 0 & 1 & 0 & -1 \\ 0 & 0 & 0 & 0 \\ 0 & -1 & 0 & 1 \end{bmatrix} \begin{Bmatrix} u_4 \\ v_4 \\ u_3 \\ v_3 \end{Bmatrix} \quad (5\text{-}17)$$

(f) Member 6

$$\begin{Bmatrix} U_1 \\ V_1 \\ U_4 \\ V_4 \end{Bmatrix} = \frac{EA}{L} \begin{bmatrix} 1 & 0 & -1 & 0 \\ 0 & 0 & 0 & 0 \\ -1 & 0 & 1 & 0 \\ 0 & 0 & 0 & 0 \end{bmatrix} \begin{Bmatrix} u_1 \\ v_1 \\ u_4 \\ v_4 \end{Bmatrix} \quad (5\text{-}18)$$

3   The superposition of the matrix quantities defined by expressions (5-12), (5-14), (5-15), (5-16), (5-17), and (5-18) will be performed, using the logic described by expression (4-82). This will develop the general force-displacement relation for the analysis of the plane truss problem. The total stiffness matrix of the truss is an 8 × 8 matrix; however, it will be reduced to a 5 × 5 matrix when the prescribed boundary conditions will be introduced into the analysis. The total stiffness matrix in general notation is given by expression (5-19) in which the superscripts define the member numbers. The numerical form of the same matrix is given by expression (5-20). Due to the support conditions of the truss, the displacements u_1, v_1, and v_4 are all zero. These boundary conditions will reduce the matrix expression (5-20) into a simpler form.

$$\{P\} = \frac{EA}{L}$$

$$\times \begin{bmatrix} \bar{K}^1_{11}+\bar{K}^4_{11}+\bar{K}^6_{11} & \bar{K}^1_{12} & \bar{K}^4_{13} & \bar{K}^6_{14} \\ \bar{K}^1_{21} & \bar{K}^1_{22}+\bar{K}^2_{22}+\bar{K}^3_{22} & \bar{K}^2_{23} & \bar{K}^3_{24} \\ \bar{K}^4_{31} & \bar{K}^2_{32} & \bar{K}^2_{33}+\bar{K}^4_{33}+\bar{K}^5_{33} & \bar{K}^5_{34} \\ \bar{K}^6_{41} & \bar{K}^3_{42} & \bar{K}^5_{43} & \bar{K}^3_{44}+\bar{K}^5_{44}+\bar{K}^6_{44} \end{bmatrix} \begin{Bmatrix} u_1 \\ v_1 \\ u_2 \\ v_2 \\ u_3 \\ v_3 \\ u_4 \\ v_4 \end{Bmatrix} \quad (5\text{-}19)$$

$$\{P\} = \frac{EA}{L}$$

$$\times \begin{bmatrix} 1.36 & 0.48 & 0 & 0 & -0.36 & -0.48 & -1 & 0 \\ 0.48 & 1.64 & 0 & -1 & -0.48 & -0.64 & 0 & 0 \\ 0 & 0 & 1.36 & -0.48 & -1 & 0 & -0.36 & 0.48 \\ 0 & -1 & -0.48 & 1.64 & 0 & 0 & 0.48 & -0.64 \\ -3.6 & -0.48 & -1 & 0 & 1.36 & 0.48 & 0 & 0 \\ -0.48 & -0.64 & 0 & 0 & 0.48 & 1.64 & 0 & -1 \\ -1 & 0 & -0.36 & 0.48 & 0 & 0 & 1.36 & -0.48 \\ 0 & 0 & 0.48 & -0.64 & 0 & -1 & -0.48 & 1.64 \end{bmatrix} \begin{Bmatrix} u_1 \\ v_1 \\ u_2 \\ v_2 \\ u_3 \\ v_3 \\ u_4 \\ v_4 \end{Bmatrix} \quad (5\text{-}20)$$

$$\begin{Bmatrix} U_2 \\ V_2 \\ U_3 \\ V_3 \\ U_4 \end{Bmatrix} = \frac{EA}{L} \begin{bmatrix} 1.36 & -0.48 & -1.0 & 0 & -0.36 \\ -0.48 & 1.64 & 0 & 0 & 0.48 \\ -1.0 & 0 & 1.36 & 0.48 & 0 \\ 0 & 0 & 0.48 & 1.64 & 0 \\ -0.36 & 0.48 & 0 & 0 & 1.36 \end{bmatrix} \begin{Bmatrix} u_2 \\ v_2 \\ u_3 \\ v_3 \\ u_4 \end{Bmatrix} \text{kips} \quad (5\text{-}21)$$

4   The externally applied loads $\{Q\}$ at the nodes of the truss have components in the X and Y directions similar to the components of the member forces $\{P\}$. To satisfy the equilibrium conditions at the nodes, the sum of the components of the forces from each member joining at the same node must be equal to the corresponding components of the externally applied nodal loads. This can be mathematically expressed by Equation (5-22).

$$\{Q\}_{\text{node}} = \Sigma \{P\}_{\text{node}} \quad (5\text{-}22)$$

The externally applied loads to our truss problem is given by the matrix

$$\{Q\} = \begin{Bmatrix} 0 \\ 0 \\ 30 \\ 0 \\ 0 \end{Bmatrix} \text{ kips} \tag{5-23}$$

The unknown nodal displacements $\{\delta\}$ are computed by rearranging expression (5-21) and substituting (5-23) as the actual applied loads with the justification of Equation (5-22).

$$\begin{Bmatrix} u_2 \\ v_2 \\ u_3 \\ v_3 \\ u_4 \end{Bmatrix} = \frac{L}{AE} \begin{bmatrix} 1.36 & -0.48 & -1.0 & 0 & -0.36 \\ -0.48 & 1.64 & 0 & 0 & 0.48 \\ -1.0 & 0 & 1.36 & 0.48 & 0 \\ 0 & 0 & 0.48 & 1.64 & 0 \\ -0.36 & 0.48 & 0 & 0 & 1.36 \end{bmatrix}^{-1} \begin{Bmatrix} 0 \\ 0 \\ 30 \\ 0 \\ 0 \end{Bmatrix} \tag{5-24}$$

The matrix operations of Equation (5-24) are performed by the aid of a digital computer. The results are given below.

$$\{\delta\} = \begin{Bmatrix} u_2 \\ v_2 \\ u_3 \\ v_3 \\ u_4 \end{Bmatrix} = \frac{L}{AE} \begin{Bmatrix} 0.6833337 \times 10^2 \\ 0.1640000 \times 10^2 \\ 0.8063340 \times 10^2 \\ -0.2360002 \times 10^2 \\ 0.1230003 \times 10^2 \end{Bmatrix} \text{ in.} \tag{5-25}$$

5 To compute directly the axial member forces, the axial elongations of the members are multiplied by (AE/L). Let us call P_1, P_2, P_3, P_4, P_5, and P_6 the axial member forces. Then we can write the relation

$$P_1 = \frac{AE}{L}\{(u_2 - u_1)\cos\alpha + (v_2 - v_1)\sin\alpha\}$$

$$P_2 = \frac{AE}{L}\{(u_3 - u_2)\cos\alpha + (v_3 - v_2)\sin\alpha\}$$

$$P_3 = \frac{AE}{L}\{(u_4 - u_3)\cos\alpha + (v_4 - v_3)\sin\alpha\}$$

$$\tag{5-26}$$

$$P_4 = \frac{AE}{L}\{(u_3 - u_4)\cos\alpha + (v_3 - v_1)\sin\alpha\}$$

$$P_5 = \frac{AE}{L}\{(u_4 - u_3)\cos\alpha + (v_4 - v_3)\sin\alpha\}$$

$$P_6 = \frac{AE}{L}\{(u_4 - u_1)\cos\alpha + (v_4 - v_1)\sin\alpha\}$$

The numerical solution of Equations (5-26) using the displacement values from Equation (5-25) yields the axial member forces;

$$\begin{Bmatrix} P_1 \\ P_2 \\ P_3 \\ P_4 \\ P_5 \\ P_6 \end{Bmatrix} = \begin{Bmatrix} 16.4 \\ 12.3 \\ -20.5 \\ 29.5 \\ -23.6 \\ 12.3 \end{Bmatrix} \cdot \frac{AE}{L} \text{ kips} \qquad (5\text{-}27)$$

The positive sign denotes tensile member forces.

Example 5-2

Perform the structural analysis of the given thin plate, Fig. 5-5, using the finite-element method. The plate is 5 ft wide, 6 ft long and 1.5 in.

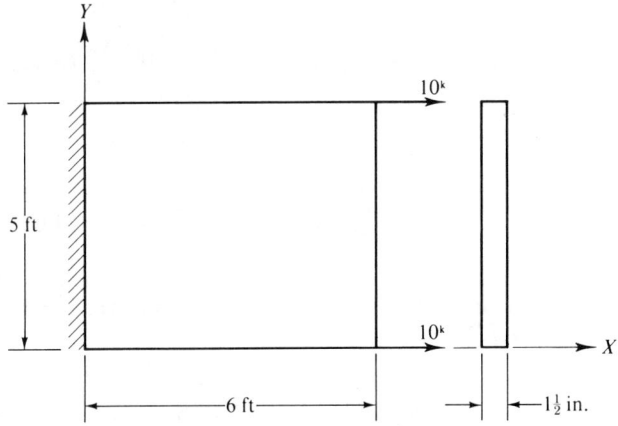

Figure 5-5

thick. It is fixed along one short side. It is acting under pure tensile loads, 10 kips each, applied at two corners. Let Young's modulus be 29,000 ksi and Poisson's ratio 0.3.

Solution

The analysis of this problem will follow the general logic described in the first part of this chapter.

1. The idealization of the structure is shown in Fig. 5-6. Four triangular elements are used for the mesh generation. The node points and elements are numbered as noted in Fig. 5-6.

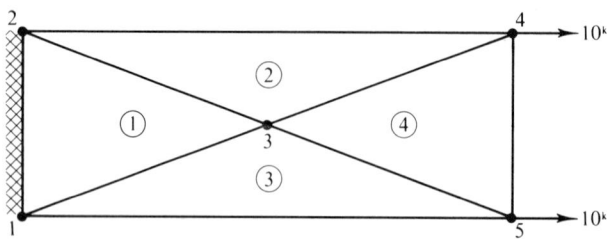

Figure 5-6

2. The geometric dimensions and elastic properties of the plate are described in the definition of the problem.
3. The boundary condition is defined by the fixity of side 1–2. Therefore, the displacements of the nodes 1 and 2 are zero.
4. This is a plane stress analysis problem. The elements used are constant-strain triangles.
5. The governing matrix equation for the analysis of this problem is given by

$$[K]\{\delta\} = \{Q\} \qquad (4\text{-}71)$$

where the stiffness matrix $[k]$ is defined by Equation (4-69).

$$[K] = \sum_{1}^{n} [k] \qquad (4\text{-}69)$$

The expression which defines the element stiffness matrix $[k]$ is given by expression (4-73).

$$[k] = \{B\}^T [D] \{B\} \cdot t \cdot \frac{dx \cdot dy}{2} \qquad (4\text{-}73)$$

The displacement-strain matrix $\{B\}$ is defined by Equations 4-10, 4-34, and 4-36. For a triangular element defined by the nodes i,

j, and m, the general matrix $\{B\}$ is written as

$$\{B\} = \frac{1}{2A} \begin{Bmatrix} y_j - y_m & 0 & y_m - y_i & 0 & y_i - y_j & 0 \\ 0 & x_m - x_j & 0 & x_i - x_m & 0 & x_j - x_i \\ x_m - x_j & y_j - y_m & x_i - x_m & y_m - y_i & x_j - x_i & y_i - y_j \end{Bmatrix}$$

(5-28)

The elasticity matrix for plane stress analysis problems is given by expression (4-44).

$$\{D\} = \frac{E}{1 - \nu^2} \begin{bmatrix} 1 & \nu & 0 \\ \nu & 1 & 0 \\ 0 & 0 & \frac{1 - \nu}{2} \end{bmatrix} \qquad (4\text{-}44)$$

The thickness of the plate is defined by t and the area of the element is given by A.

The general nodal displacement matrix defined by expression (4-1), can be written for this particular problem as follows:

$$\{\delta\} = \begin{Bmatrix} u_1 \\ v_1 \\ u_2 \\ v_2 \\ u_3 \\ v_3 \\ u_4 \\ v_4 \\ u_5 \\ v_5 \end{Bmatrix}$$

The external loads are applied only at the nodes 4 and 5 and are illustrated by the following matrix.

$$\{Q\} = \begin{Bmatrix} 0 \\ 0 \\ 0 \\ 0 \\ 0 \\ 0 \\ 10 \\ 0 \\ 10 \\ 0 \end{Bmatrix} \text{ kips}$$

Generation of Element Stiffness Matrices

To generate the element stiffness matrix, the general expression (4-73) is used:

$$[K] = \{B\}^T[D]\{B\}t \cdot A$$

The coordinates of the nodal points are given in Fig. 5-7.

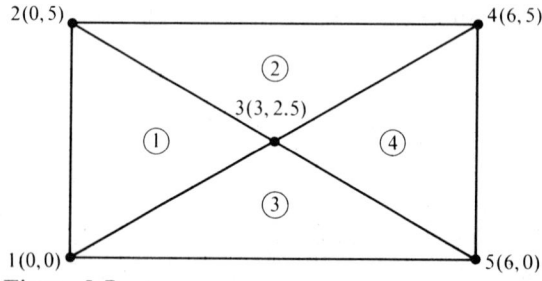

Figure 5-7

The topology of the elements uses counterclockwise direction.

Element 1

(a) The displacement-strain matrix $\{B\}$ is developed by the use of the general expression (5-28). This element is defined by the nodes 1, 3, and 2.

$$\{B\} = \frac{12}{2 \cdot \frac{15}{2}(144)} \begin{Bmatrix} (2.5-5) & 0 & 5 & 0 & -2.5 & 0 \\ 0 & -3 & 0 & 0 & 0 & 3 \\ -3 & (2.5-5) & 0 & 5 & 3 & -2.5 \end{Bmatrix}$$

which simplifies into the following matrix form:

$$\{B\} = \frac{1}{180} \begin{Bmatrix} -2.5 & 0 & 5 & 0 & -2.5 & 0 \\ 0 & -3 & 0 & 0 & 0 & 3 \\ -3 & -2.5 & 0 & 5 & 3 & -2.5 \end{Bmatrix}$$

(b) The elasticity matrix $[D]$ is calculated by the use of expression (4-44).

$$[D] = \frac{29000}{1 - (0.3)^2} \begin{bmatrix} 1 & 0.3 & 0 \\ 0.3 & 1 & 0 \\ 0 & 0 & \frac{1 - 0.3}{2} \end{bmatrix}$$

which simplifies into the following matrix form:

$$[D] = \frac{29000}{0.91} \begin{bmatrix} 1 & 0.3 & 0 \\ 0.3 & 1 & 0 \\ 0 & 0 & 0.35 \end{bmatrix}$$

(c) The calculation of the element stiffness matrix:

$$[K]_1 = \{B\}^T [D] \{B\} t \cdot A$$

The constants of the matrix multiplication are precomputed:

$$\left\{ \left(\frac{1}{180}\right)^2 \left(\frac{29000}{0.91}\right)(1.5) \left(\frac{5 \times 3 \times 144}{2}\right) \right\} = 1593.4$$

$[K]_1 = (1593.4)$

$$\begin{bmatrix} -2.5 & 0 & -3 \\ 0 & -3 & -2.5 \\ 5 & 0 & 0 \\ 0 & 0 & 5 \\ -2.5 & 0 & 3 \\ 0 & 3 & -2.5 \end{bmatrix} \begin{bmatrix} 1 & 0.3 & 0 \\ 0.3 & 1 & 0 \\ 0 & 0 & 0.35 \end{bmatrix} \begin{bmatrix} -2.5 & 0 & 5 & 0 & -2.5 & 0 \\ 0 & -3 & 0 & 0 & 0 & 3 \\ -3 & -2.5 & 0 & 5 & 3 & -2.5 \end{bmatrix}$$

$[K]_1 = 1593.4$

$$\times \begin{bmatrix} -2.5 & -0.75 & -1.05 \\ -0.9 & -3 & -0.875 \\ 5 & 1.5 & 0 \\ 0 & 0 & 1.75 \\ -2.5 & -0.75 & 1.05 \\ 0.9 & 3 & -0.875 \end{bmatrix} \begin{bmatrix} -2.5 & 0 & 5 & 0 & -2.5 & 0 \\ 0 & -3 & 0 & 0 & 0 & 3 \\ -3 & -2.5 & 0 & 5 & 3 & -2.5 \end{bmatrix}$$

$[K]_1 = 1593.4$

$$\times \begin{bmatrix} 9.40 & 4.875 & | & -12.5 & -5.25 & | & 3.10 & 0.375 \\ & (K_{11}) & | & & (K_{13}) & | & & (K_{12}) \\ 4.875 & 11.19 & | & -4.5 & -4.375 & | & -0.375 & -6.81 \\ \hline -12.5 & -4.5 & | & 25 & 0 & | & -12.5 & 4.50 \\ & (K_{31}) & | & & (K_{33}) & | & & (K_{32}) \\ -5.25 & -4.375 & | & 0 & 8.75 & | & 5.25 & -4.375 \\ \hline 3.10 & -0.375 & | & -12.5 & 5.25 & | & 9.40 & -4.875 \\ & (K_{21}) & | & & (K_{23}) & | & & (K_{22}) \\ 0.375 & -6.81 & | & 4.50 & -4.375 & | & -4.875 & 11.19 \end{bmatrix}$$

Element 2

This element is defined by the nodes 2, 3, and 4. The matrix quantities for the elements 2, 3, and 4 will be given numerically since they are defined for the development of element 1's stiffness matrix.

(a) The displacement-strain matrix $\{B\}$.

$$\{B\} = \frac{1}{180} \begin{Bmatrix} -2.5 & 0 & 0 & 0 & 2.5 & 0 \\ 0 & 3 & 0 & -6 & 0 & 3 \\ 3 & -2.5 & -6 & 0 & 3 & 2.5 \end{Bmatrix}$$

(b) The elasticity matrix $[D]$.

$$[D] = \frac{29000}{0.91} \begin{bmatrix} 1 & 0.3 & 0 \\ 0.3 & 1 & 0 \\ 0 & 0 & 0.35 \end{bmatrix}$$

(c) The element stiffness matrix $[K]_2$.

$$[K]_2 = 1593.4 \times \begin{bmatrix} -2.5 & 0 & 3 \\ 0 & 3 & -2.5 \\ 0 & 0 & -6 \\ 0 & -6 & 0 \\ 2.5 & 0 & 3 \\ 0 & 3 & 2.5 \end{bmatrix} \begin{bmatrix} 1 & 0.3 & 0 \\ 0.3 & 1 & 0 \\ 0 & 0 & 0.35 \end{bmatrix} \begin{bmatrix} -2.5 & 0 & 0 & 0 & 2.5 & 0 \\ 0 & 3 & 0 & -6 & 0 & 3 \\ 3 & -2.5 & -6 & 0 & 3 & 2.5 \end{bmatrix}$$

$$[K]_2 = 1593.4 \times \begin{bmatrix} -2.5 & -0.75 & 1.05 \\ 0.9 & 3 & -0.875 \\ 0 & 0 & -2.10 \\ -1.8 & -6 & 0 \\ 2.5 & 0.75 & 1.05 \\ 0.9 & 3 & 0.875 \end{bmatrix} \begin{bmatrix} -2.5 & 0 & 0 & 0 & 2.5 & 0 \\ 0 & 3 & 0 & -6 & 0 & 3 \\ 3 & -2.5 & -6 & 0 & 3 & 2.5 \end{bmatrix}$$

$$[K]_2 = 1593.4 \times \begin{bmatrix} 9.40 & -4.875 & -6.3 & 4.5 & -3.1 & 0.375 \\ & (K_{22}) & & (K_{23}) & & (K_{24}) \\ -4.875 & 11.19 & 5.25 & -18 & -0.375 & 6.81 \\ \hline -6.3 & 5.25 & 12.60 & 0 & -6.30 & -5.25 \\ & (K_{32}) & & (K_{33}) & & (K_{34}) \\ 4.5 & -18 & 0 & 36 & -4.5 & -18 \\ \hline -3.1 & -0.375 & -6.30 & -4.5 & 9.4 & 4.875 \\ & (K_{42}) & & (K_{43}) & & (K_{44}) \\ 0.375 & 6.81 & -5.25 & -18 & 4.875 & 11.19 \end{bmatrix}$$

Element 3

This element is defined by the nodes 1, 5, and 3.

(a) The displacement-strain matrix, $\{B\}$.

$$\{B\} = \frac{1}{180} \begin{bmatrix} -2.5 & 0 & 2.5 & 0 & 0 & 0 \\ 0 & -3 & 0 & -3 & 0 & 6 \\ -3 & -2.5 & -3 & 2.5 & 6 & 0 \end{bmatrix}$$

(b) The elasticity matrix $[D]$ will be the same for all elements:

$$[D] = \frac{29000}{0.91} \begin{bmatrix} 1 & 0.3 & 0 \\ 0.3 & 1 & 0 \\ 0 & 0 & 0.35 \end{bmatrix}$$

(c) The element stiffness matrix $[K]_3$.

$$[K]_3 = 1593.4$$

$$\times \begin{bmatrix} -2.5 & 0 & -3 \\ 0 & -3 & -2.5 \\ 2.5 & 0 & -3 \\ 0 & -3 & 2.5 \\ 0 & 0 & 6 \\ 0 & 6 & 0 \end{bmatrix} \begin{bmatrix} 1 & 0.3 & 0 \\ 0.3 & 1 & 0 \\ 0 & 0 & 0.35 \end{bmatrix} \begin{bmatrix} -2.5 & 0 & 2.5 & 0 & 0 & 0 \\ 0 & -3 & 0 & -3 & 0 & 6 \\ -3 & -2.5 & -3 & 2.5 & 6 & 0 \end{bmatrix}$$

$$[K]_3 = 1593.4$$

$$\times \begin{bmatrix} -2.5 & -0.75 & -1.05 \\ -0.9 & -3 & -0.875 \\ 2.5 & 0.75 & -1.05 \\ -0.9 & -3 & -0.875 \\ 0 & 0 & 2.10 \\ 1.8 & 6 & 0 \end{bmatrix} \begin{bmatrix} -2.5 & 0 & 2.5 & 0 & 0 & 0 \\ 0 & -3 & 0 & -3 & 0 & 6 \\ -3 & -2.5 & -3 & 2.5 & 6 & 0 \end{bmatrix}$$

$$[K]_3 = 1593.4 \times \begin{bmatrix} 9.40 & 4.875 & | & -3.10 & -0.375 & | & -6.30 & -4.5 \\ & (K_{11}) & | & & (K_{15}) & | & & (K_{13}) \\ 4.875 & 11.19 & | & 0.375 & 6.81 & | & -5.25 & -18 \\ \hline -3.10 & 0.375 & | & 9.40 & -4.875 & | & -6.30 & 4.5 \\ & (K_{51}) & | & & (K_{55}) & | & & (K_{53}) \\ -0.375 & 6.81 & | & -4.875 & 11.19 & | & 5.25 & -18 \\ \hline -6.30 & -5.25 & | & -6.30 & 5.25 & | & 12.60 & 0 \\ & (K_{31}) & | & & (K_{35}) & | & & (K_{33}) \\ -4.50 & -18 & | & 4.50 & -18 & | & 0 & 36 \end{bmatrix}$$

Element 4

This element is defined by the nodes 5, 4, and 3.

(a) The displacement-strain matrix $\{B\}$.

$$\{B\} = \frac{1}{180} \begin{bmatrix} 2.5 & 0 & 2.5 & 0 & -5 & 0 \\ 0 & -3 & 0 & 3 & 0 & 0 \\ -3 & 2.5 & 3 & 2.5 & 0 & -5 \end{bmatrix}$$

(b) The elasticity matrix $[D]$.

$$[D] = \frac{29000}{0.91} \begin{bmatrix} 1 & 0.3 & 0 \\ 0.3 & 1 & 0 \\ 0 & 0 & 0.35 \end{bmatrix}$$

(c) The element stiffness matrix $[K]_4$.

$$[K]_4 = 1593.4 \times \begin{bmatrix} 2.5 & 0 & -3 \\ 0 & -3 & 2.5 \\ 2.5 & 0 & 3 \\ 0 & 3 & 2.5 \\ -5 & 0 & 0 \\ 0 & 0 & -5 \end{bmatrix} \begin{bmatrix} 1 & 0.3 & 0 \\ 0.3 & 1 & 0 \\ 0 & 0 & 0.35 \end{bmatrix} \begin{bmatrix} 2.5 & 0 & 2.5 & 0 & -5 & 0 \\ 0 & -3 & 0 & 3 & 0 & 0 \\ -3 & 2.5 & 3 & 2.5 & 0 & -5 \end{bmatrix}$$

$[K]_4 = 1593.4$

$$\times \begin{bmatrix} 2.5 & 0.75 & -1.05 \\ -0.9 & -3 & 0.875 \\ 2.5 & 0.75 & 1.05 \\ 0.9 & 3 & 0.875 \\ -5 & -1.5 & 0 \\ 0 & 0 & -1.75 \end{bmatrix} \begin{bmatrix} 2.5 & 0 & 2.5 & 0 & -5 & 0 \\ 0 & -3 & 0 & 3 & 0 & 0 \\ -3 & 2.5 & 3 & 2.5 & 0 & -5 \end{bmatrix}$$

$[K]_4 = 1593.4$

$$\times \begin{bmatrix} 9.40 & -4.875 & | & 3.10 & -0.375 & | & -12.5 & 5.25 \\ (K_{55}) & & | & (K_{54}) & & | & (K_{53}) & \\ -4.875 & 11.19 & | & 0.375 & -6.81 & | & 4.5 & -4.375 \\ \hline 3.10 & 0.375 & | & 9.40 & 4.875 & | & -12.5 & -5.25 \\ (K_{45}) & & | & (K_{44}) & & | & (K_{43}) & \\ -0.375 & -6.81 & | & 4.875 & 11.19 & | & -4.5 & -4.375 \\ \hline -12.5 & 4.5 & | & -12.5 & -4.5 & | & 25 & 0 \\ (K_{35}) & & | & (K_{34}) & & | & (K_{33}) & \\ 5.25 & -4.375 & | & -5.25 & -4.375 & | & 0 & 8.75 \end{bmatrix}$$

6 The total stiffness matrix for the plate is obtained by the superposition of the element stiffness matrices. The general form of the total matrix will be in the following form.

$$[K]_{total} = \begin{array}{c} \\ 1 \\ 2 \\ 3 \\ 4 \\ 5 \end{array} \begin{bmatrix} \Sigma K_{11} & \Sigma K_{12} & \Sigma K_{13} & \Sigma K_{14} & \Sigma K_{15} \\ \Sigma K_{21} & \Sigma K_{22} & \Sigma K_{23} & \Sigma K_{24} & \Sigma K_{25} \\ \Sigma K_{31} & \Sigma K_{32} & \Sigma K_{33} & \Sigma K_{34} & \Sigma K_{35} \\ \Sigma K_{41} & \Sigma K_{42} & \Sigma K_{43} & \Sigma K_{44} & \Sigma K_{45} \\ \Sigma K_{51} & \Sigma K_{52} & \Sigma K_{53} & \Sigma K_{54} & \Sigma K_{55} \end{bmatrix}$$

$$\begin{array}{ccccc} 1 & 2 & 3 & 4 & 5 \end{array}$$

In numerical form, this matrix is

$$[K]_{\text{total}} = 1593.4 \times \begin{bmatrix}
18.8 & 9.75 & 3.1 & 0.375 & -18.8 & -9.75 & 0 & 0 & -3.1 & -0.375 \\
9.75 & 22.375 & -0.375 & -6.81 & -9.75 & -22.375 & 0 & 0 & 0.375 & 6.81 \\
3.1 & -0.375 & 18.8 & -9.75 & -18.8 & 9.75 & -3.1 & 0.375 & 0 & 0 \\
0.375 & -6.81 & -9.75 & 22.375 & 9.75 & -22.375 & -0.375 & 6.81 & 0 & 0 \\
-18.8 & -9.75 & -18.8 & 9.75 & 75.2 & 0 & -18.8 & -9.75 & -18.8 & 9.75 \\
-9.75 & -22.375 & 9.75 & -22.375 & 0 & 89.5 & -9.75 & -22.375 & 9.75 & -22.375 \\
0 & 0 & -3.1 & -0.375 & -18.8 & -9.7 & 18.8 & 9.75 & 3.1 & 0.375 \\
0 & 0 & 0.375 & 6.81 & -9.75 & -22.375 & 9.75 & 22.375 & -0.375 & -6.81 \\
-3.1 & 0.375 & 0 & 0 & -18.8 & 9.75 & 3.1 & -0.375 & 18.8 & -9.5 \\
-0.375 & 6.81 & 0 & 0 & 9.75 & -22.375 & 0.375 & -6.81 & -9.75 & 22.375
\end{bmatrix}$$

7. The computation of the nodal displacements $\{\delta\}$ is based on the general equation (4-72) which has the form

$$\{\delta\} = [K]^{-1}\{Q\} \qquad (4\text{-}72)$$

The displacement matrix $\{\delta\}$ has ten possible elements:

$$\{\delta\} = \begin{Bmatrix} u_1 \\ v_1 \\ u_2 \\ v_2 \\ u_3 \\ v_3 \\ u_4 \\ v_4 \\ u_5 \\ v_5 \end{Bmatrix}$$

in which the elements u_1, v_1, u_2, and v_2 are zero, due to the fixed boundary conditions at nodes 1 and 2. This will reduce the stiffness matrix $[K]$ to a 6×6 matrix. The remaining unknown displacements are computed by the use of Equation (4-72). The inversion of the stiffness matrix is performed by the aid of a digital computer.

$$\{\delta\} = \begin{Bmatrix} u_3 \\ v_3 \\ u_4 \\ v_4 \\ u_5 \\ v_5 \end{Bmatrix} = \left(\frac{1}{1593.4}\right) \cdot [K]^{-1} \begin{Bmatrix} 0 \\ 0 \\ 10 \\ 0 \\ 10 \\ 0 \end{Bmatrix}$$

where

$$[K]^{-1} = \begin{bmatrix} 0.0234 & 0.0000 & 0.0194 & 0.0016 & 0.0194 & -0.0016 \\ 0.0000 & 0.0498 & -0.0263 & 0.0887 & 0.0263 & 0.0887 \\ 0.0194 & -0.0263 & 0.1113 & -0.0897 & -0.0261 & -0.0753 \\ 0.0016 & 0.0887 & -0.0897 & 0.2332 & 0.0753 & 0.1933 \\ 0.0194 & 0.0263 & -0.0261 & 0.0753 & 0.1113 & 0.0897 \\ -0.0016 & 0.0887 & -0.0753 & 0.1933 & 0.0897 & 0.2332 \end{bmatrix}$$

The numerical values of the nodal displacements are

$$\{\delta\} = \begin{Bmatrix} u_3 \\ v_3 \\ u_4 \\ v_4 \\ u_5 \\ v_5 \end{Bmatrix} = \begin{Bmatrix} 0.3886 \\ 0.0000 \\ 0.8518 \\ -0.1438 \\ 0.8518 \\ 0.1438 \end{Bmatrix} \cdot \frac{1}{1593.4} \text{ in.}$$

8   The calculation of the element stresses are performed by the use of Equations (4-36) and (4-39) which relate stresses to displacements.

$$\{\sigma\} = [D]\{B\}\{\delta\}$$

Since all the values of the matrices are located at the right-hand side of the above equation, the stresses are readily obtained. The computations are done for each element.

Element 1

$$\begin{bmatrix} \sigma_x \\ \sigma_y \\ \tau_{xy} \end{bmatrix} = (0.111)$$

$$\times \begin{bmatrix} 1 & 0.3 & 0 \\ 0.3 & 1 & 0 \\ 0 & 0 & 0.35 \end{bmatrix} \begin{bmatrix} -2.5 & 0 & 5 & 0 & -2.5 & 0 \\ 0 & -3 & 0 & 0 & 0 & 3 \\ -3 & -2.5 & 0 & 5 & 3 & -2.5 \end{bmatrix} \begin{bmatrix} 0 \\ 0 \\ 0.3886 \\ 0 \\ 0 \\ 0 \end{bmatrix}$$

$$\begin{bmatrix} \sigma_x \\ \sigma_y \\ \tau_{xy} \end{bmatrix} = 0.111 \begin{bmatrix} 1 & 0.3 & 0 \\ 0.3 & 1 & 0 \\ 0 & 0 & 0.35 \end{bmatrix} \begin{bmatrix} 1.943 \\ 0 \\ 0 \end{bmatrix} = 0.111 \begin{bmatrix} 1.943 \\ 0.5829 \\ 0 \end{bmatrix} \text{ksi}$$

$$\begin{Bmatrix} \sigma_x \\ \sigma_y \\ \tau_{xy} \end{Bmatrix} = \begin{Bmatrix} 0.2160 \\ 0.0646 \\ 0 \end{Bmatrix} \text{ksi}$$

The principal stresses are also readily computed using expressions (3-10):

$$\sigma_{max} = \frac{\sigma_x + \sigma_y}{2} + \sqrt{\left(\frac{\sigma_x - \sigma_y}{2}\right)^2 + \tau_{xy}^2} = 0.216 \text{ ksi}$$

$$\sigma_{min} = \frac{\sigma_x + \sigma_y}{2} - \sqrt{\left(\frac{\sigma_x - \sigma_y}{2}\right)^2 + \tau_{xy}^2} = 0.0646 \text{ ksi}$$

Element 2

$$\begin{Bmatrix} \sigma_x \\ \sigma_y \\ \tau_{xy} \end{Bmatrix} = (0.111)$$

$$\times \begin{bmatrix} 1 & 0.3 & 0 \\ 0.3 & 1 & 0 \\ 0 & 0 & 0.35 \end{bmatrix} \begin{bmatrix} -2.5 & 0 & 0 & 0 & 2.5 & 0 \\ 0 & 3 & 0 & -6 & 0 & 3 \\ 3 & -2.5 & -6 & 0 & 3 & 2.5 \end{bmatrix} \begin{bmatrix} 0 \\ 0 \\ 0.3886 \\ 0 \\ 0.8518 \\ -0.1438 \end{bmatrix}$$

$$\begin{Bmatrix} \sigma_x \\ \sigma_y \\ \tau_{xy} \end{Bmatrix} = 0.111 \cdot \begin{bmatrix} 1 & 0.3 & 0 \\ 0.3 & 1 & 0 \\ 0 & 0 & 0.35 \end{bmatrix} \begin{bmatrix} 2.13 \\ -0.431 \\ -0.1357 \end{bmatrix} = 0.111 \begin{bmatrix} 2.0 \\ 0.208 \\ -0.0475 \end{bmatrix}$$

$$\begin{Bmatrix} \sigma_x \\ \sigma_y \\ \tau_{xy} \end{Bmatrix} = \begin{Bmatrix} 0.22200 \\ 0.02310 \\ -0.00527 \end{Bmatrix} \text{ksi}$$

The principal stresses for element 2 are given below which are computed

by expression (3-10):
$$\sigma_{max} = 0.22225 \text{ ksi}$$
$$\sigma_{min} = 0.02285 \text{ ksi}$$

Element 3

$$\begin{Bmatrix} \sigma_x \\ \sigma_y \\ \tau_{xy} \end{Bmatrix} = (0.111) \times \begin{bmatrix} 1 & 0.3 & 0 \\ 0.3 & 1 & 0 \\ 0 & 0 & 0.35 \end{bmatrix} \begin{bmatrix} -2.5 & 0 & 2.5 & 0 & 0 & 0 \\ 0 & -3 & 0 & -3 & 0 & 6 \\ -3 & -2.5 & -3 & 2.5 & 6 & 0 \end{bmatrix} \begin{bmatrix} 0 \\ 0 \\ 0.8518 \\ 0.1438 \\ 0.3886 \\ 0 \end{bmatrix}$$

$$\begin{Bmatrix} \sigma_x \\ \sigma_y \\ \tau_{xy} \end{Bmatrix} = (0.111) \begin{bmatrix} 1 & 0.3 & 0 \\ 0.3 & 1 & 0 \\ 0 & 0 & 0.35 \end{bmatrix} \begin{bmatrix} 2.13 \\ -0.4314 \\ 0.1345 \end{bmatrix} = 0.111 \begin{bmatrix} 2.0 \\ 0.2076 \\ 0.0471 \end{bmatrix}$$

$$\begin{Bmatrix} \sigma_x \\ \sigma_y \\ \tau_{xy} \end{Bmatrix} = \begin{Bmatrix} 0.2220 \\ 0.0231 \\ 0.0053 \end{Bmatrix} \text{ ksi}$$

The principal stresses for element 3 are computed by expression (3-10):
$$\sigma_{max} = 0.22225 \text{ ksi}$$
$$\sigma_{min} = 0.02285 \text{ ksi}$$

Element 4

$$\begin{Bmatrix} \sigma_x \\ \sigma_y \\ \tau_{xy} \end{Bmatrix} = (0.111) \times \begin{bmatrix} 1 & 0.3 & 0 \\ 0.3 & 1 & 0 \\ 0 & 0 & 0.35 \end{bmatrix} \begin{bmatrix} 2.5 & 0 & 2.5 & 0 & -5 & 0 \\ 0 & -3 & 0 & 3 & 0 & 0 \\ -3 & 2.5 & 3 & 2.5 & 0 & -5 \end{bmatrix} \begin{bmatrix} 0.8518 \\ 0.1438 \\ 0.8518 \\ -0.1438 \\ 0.3886 \\ .0 \end{bmatrix}$$

$$\begin{Bmatrix} \sigma_x \\ \sigma_y \\ \tau_{xy} \end{Bmatrix} = (0.111) \begin{bmatrix} 1 & 0.3 & 0 \\ 0.3 & 1 & 0 \\ 0 & 0 & 0.35 \end{bmatrix} \begin{bmatrix} 2.316 \\ -0.8628 \\ 0 \end{bmatrix} = 0.111 \begin{bmatrix} 2.0572 \\ -0.168 \\ 0 \end{bmatrix}$$

$$\begin{Bmatrix} \sigma_x \\ \sigma_y \\ \tau_{xy} \end{Bmatrix} = \begin{Bmatrix} 0.22800 \\ -0.01864 \\ 0 \end{Bmatrix} \text{ ksi}$$

The principal stresses for element-4 are computed by expression (3-10):

$$\sigma_{max} = 0.2280 \text{ ksi}$$
$$\sigma_{min} = -0.01864 \text{ ksi}$$

Example 5-3

Perform the analysis of the simply supported beam shown in Fig. 5-8. The beam spans 4 ft and supports a uniform load of 2 kips/ft. It is made

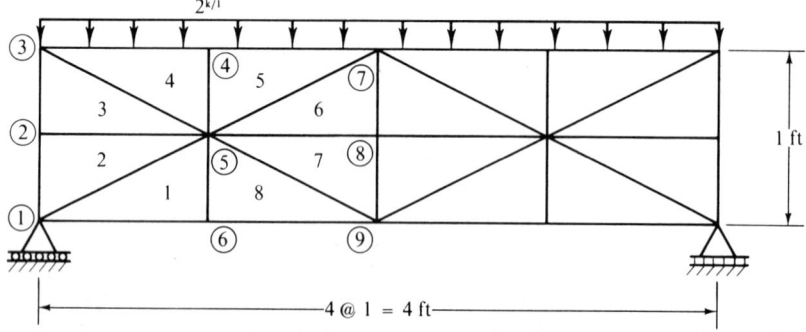

Figure 5-8

of A-36 steel whose Young's modulus is 30,000 ksi and whose Poisson ratio is 0.3. Use triangular elements for mesh generation. The thickness is equal to one inch.

Solution

The analysis of the given beam will follow the logic used in the discussion of Problem 5-2. This is an illustrative problem; its aim is a didactic one without seeking practical accuracy in the numerical solution. To satisfy the accuracy, the elements must decrease in size and increase in quantity.

1. The idealization of the beam is done by the use of sixteen triangular elements as shown in Fig. 5-9. Because of the symmetry

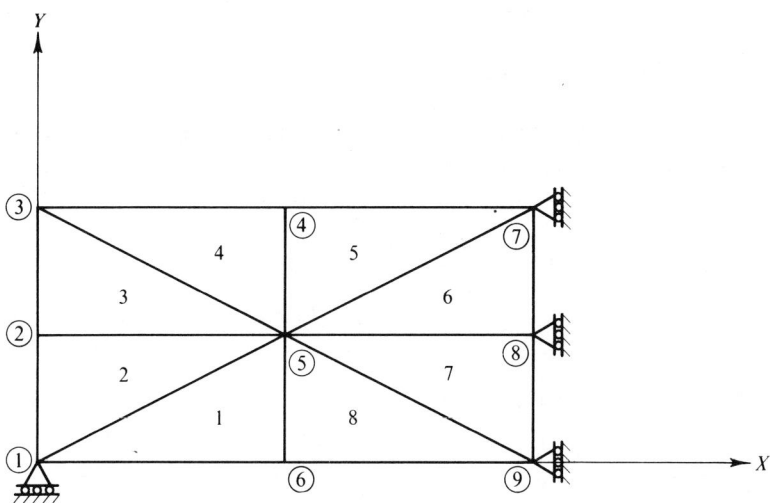

Figure 5-9

of the structure, only one-half of the beam is used during the analysis. The elements and corresponding nodes are numbered as shown in Fig. 5-9.

2. The geometric dimensions of the problem are given in Fig. 5-8 and the elastic properties are described in the definition of the problem. Because of the symmetry, only one-half of the beam is considered for the analysis.

3. The beam is supported at nodes 1, 7, 8, and 9. These boundary conditions will make the horizontal and vertical displacements of the node 1 and the horizontal displacements of nodes 7, 8, and 9 zero, since they lie on the midspan line of symmetry.

4. The approximation of the solution of the beam problem is done by the use of constant-strain triangular elements. If enough number of elements are used, the solution will tend to converge to the correct one.

5. The governing matrix equation of the finite element analysis is given by expression (4-71) which is

$$[K] \cdot \{\delta\} = \{Q\} \qquad (4\text{-}71)$$

where the total structural stiffness $[K]$ is the sum of the individual element matrices defined by Equation (4-69)

$$[K] = \sum_{1}^{n} [K] \qquad (4\text{-}69)$$

The element stiffness matrix is defined by expression (4-73) as

$$[K] = \{B\}^T[D]\{B\} \cdot t \cdot A \qquad (4\text{-}73)$$

The displacement-strain matrix $\{B\}$ is defined by the matrix expression (5-28).

The elasticity matrix related to the plane stress problem is illustrated by the matrix given by expression (4-44).

$$[D] = \frac{E}{1-\nu^2} \begin{bmatrix} 1 & \nu & 0 \\ \nu & 0 & 0 \\ 0 & 0 & \frac{1-\nu}{2} \end{bmatrix}$$

The displacement matrix $\{\delta\}$ for the half-beam is

$$\{\delta\} = \begin{Bmatrix} u_1 \\ v_1 \\ u_2 \\ v_2 \\ u_3 \\ v_3 \\ u_4 \\ v_4 \\ \vdots \\ u_9 \\ v_9 \end{Bmatrix}$$

The load matrix $\{Q\}$ defines the equivalent nodal loads due to the uniform load of 2 kips/ft over the span. Only one-half of the beam is considered due to the symmetry. All the loads are in the Y direction.

$$\{Q\} = \begin{Bmatrix} Q_3 \\ Q_4 \\ Q_7 \end{Bmatrix} = \begin{Bmatrix} -1 \\ -2 \\ -1 \end{Bmatrix} \text{ kips}$$

Generation of Element Stiffness Matrices

The basic equation (4-73) is used for the development of the element stiffness matrices:

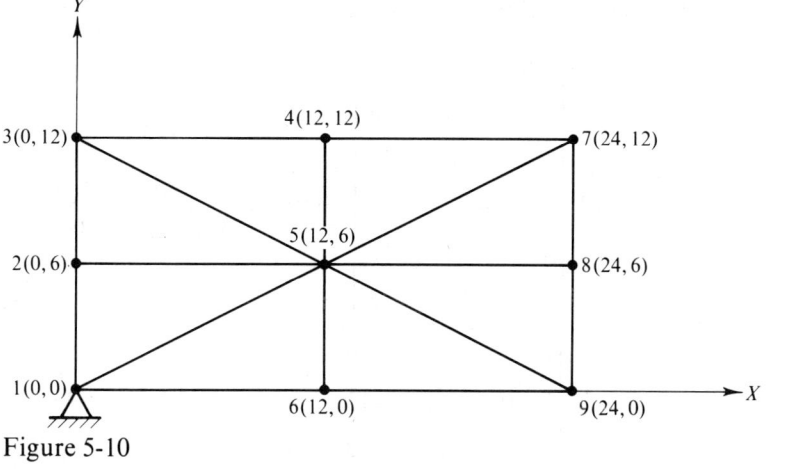

Figure 5-10

$$[K] = \{B\}^T[D]\{B\} \cdot t \cdot A \qquad (4\text{-}73)$$

The coordinates of the nodal points are given in Fig. 5-10.

Element 1

The area of all the triangles will be the same. The thickness of the elements is unity.

$$\text{area} = (12)(6)(1/2) = 36 \text{ in.}^2$$

Element 1 is defined by the nodes 1, 5, and 6. Clockwise direction is used in correlating the elements and nodes. As long as the same direction is used throughout the same problem, the solutions will be consistent.

(a) Displacement-strain matrix $\{B\}$.

$$\{B\} = \frac{1}{72} \begin{bmatrix} 6 & 0 & 0 & 0 & -6 & 0 \\ 0 & 0 & 0 & -12 & 0 & 12 \\ 0 & 6 & -12 & 0 & 12 & -6 \end{bmatrix}$$

(b) Elasticity matrix $[D]$.

$$[D] = \frac{30{,}000}{0.91} \begin{bmatrix} 1 & 0.3 & 0 \\ 0.3 & 1 & 0 \\ 0 & 0 & 0.35 \end{bmatrix}$$

(c) The element stiffness matrix $[K]$.

$$[K]_1 = \frac{1}{72} \begin{bmatrix} 6 & 0 & 0 \\ 0 & 0 & 6 \\ 0 & 0 & -12 \\ 0 & -12 & 0 \\ -6 & 0 & 12 \\ 0 & 12 & -6 \end{bmatrix} \frac{30{,}000}{0.91} \begin{bmatrix} 1 & 0.3 & 0 \\ 0.3 & 1 & 0 \\ 0 & 0 & 0.35 \end{bmatrix}$$

$$\times \frac{1}{72} \begin{bmatrix} 6 & 0 & 0 & 0 & -6 & 0 \\ 0 & 0 & 0 & -12 & 0 & 12 \\ 0 & 6 & -12 & 0 & 12 & -6 \end{bmatrix} \times 1 \times 36$$

$$[K]_1 = 8250 \begin{bmatrix} 1.0 & 0 & 0 & -0.6 & -1 & 0.6 \\ 0 & 0.35 & -0.7 & 0 & 0.7 & -0.35 \\ 0 & -0.7 & 1.4 & 0 & -1.4 & 0.7 \\ -0.6 & 0 & 0 & 4 & 0.6 & -4 \\ -1 & 0.7 & -1.4 & 0.6 & 2.4 & -1.3 \\ 0.6 & -0.35 & 0.7 & -4 & -1.3 & 4.35 \end{bmatrix}$$

Element 2

This element is defined by the nodes 1, 2, and 5.

(a) The displacement-strain matrix $\{B\}$ is computed by the use of matrix (5-28)

$$\{B\} = \frac{1}{72} \begin{bmatrix} 0 & 0 & 6 & 0 & -6 & 0 \\ 0 & 12 & 0 & -12 & 0 & 0 \\ 12 & 0 & -12 & 6 & 0 & -6 \end{bmatrix}$$

(b) The elasticity matrix $[D]$ is computed according to expression (4-44). The matrix $[D]$ will be the same for all elements since the beam is elastic, homogeneous and isotropic.

$$[D] = \frac{30{,}000}{0.91} \begin{bmatrix} 1 & 0.3 & 0 \\ 0.3 & 1 & 0 \\ 0 & 0 & 0.35 \end{bmatrix}$$

INTRODUCTION 129

(c) The element stiffness matrix $[K]$ is computed according to the formula given in Equation (4-73).

$$[K]_2 = \frac{1}{72} \begin{bmatrix} 0 & 0 & 12 \\ 0 & 12 & 0 \\ 6 & 0 & -12 \\ 0 & -12 & 6 \\ -6 & 0 & 0 \\ 0 & 0 & -6 \end{bmatrix} \frac{30{,}000}{0.91} \begin{bmatrix} 1 & 0.3 & 0 \\ 0.3 & 1 & 0 \\ 1 & 0 & 0.35 \end{bmatrix}$$

$$\times \frac{1}{72} \begin{bmatrix} 0 & 0 & 6 & 0 & -6 & 0 \\ 0 & 12 & 0 & -12 & 0 & 0 \\ 12 & 0 & -12 & 6 & 0 & -6 \end{bmatrix} 1 \times 36$$

$$\{K\}_2 = 8250 \begin{bmatrix} 1.4 & 0 & | & -1.4 & 0.7 & | & 0 & -0.7 \\ 0 & 4 & | & 0.6 & -4 & | & -0.6 & 0 \\ \hline -1.4 & 0.6 & | & 2.4 & -1.3 & | & -1 & 0.7 \\ 0.7 & -4 & | & -1.3 & 4.35 & | & 0.6 & -0.35 \\ \hline 0 & -0.6 & | & -1 & 0.6 & | & 1 & 0 \\ -0.7 & 0 & | & 0.7 & -0.35 & | & 0 & 0.35 \end{bmatrix}$$

Element 3

This element is defined by the nodes 2, 3, and 5.

(a) The displacement-strain matrix $\{B\}$.

$$\{B\} = \frac{1}{72} \begin{bmatrix} 6 & 0 & 0 & 0 & -6 & 0 \\ 0 & 12 & 0 & -12 & 0 & 0 \\ 12 & 6 & -12 & 0 & 0 & -6 \end{bmatrix}$$

(b) The elasticity matrix $[D]$.

$$[D] = \frac{30{,}000}{0.91} \cdot \begin{bmatrix} 1 & 0.3 & 0 \\ 0.3 & 1 & 0 \\ 0 & 0 & 0.35 \end{bmatrix}$$

(c) The element stiffness matrix $[K]$.

$$[K]_3 = \frac{1}{72} \begin{bmatrix} 6 & 0 & 12 \\ 0 & 12 & 6 \\ 0 & 0 & -12 \\ 0 & -12 & 0 \\ -6 & 0 & 0 \\ 0 & 0 & -6 \end{bmatrix} \frac{30{,}000}{0.91} \begin{bmatrix} 1 & 0.3 & 0 \\ 0.3 & 1 & 0 \\ 0 & 0 & 0.35 \end{bmatrix}$$

$$\cdot \frac{1}{72} \begin{bmatrix} 6 & 0 & 0 & 0 & -6 & 0 \\ 0 & 12 & 0 & -12 & 0 & 0 \\ 12 & 6 & -12 & 0 & 0 & -6 \end{bmatrix} \times 36$$

$$[K]_3 = 8250 \begin{bmatrix} 2.4 & 1.3 & -1.4 & -0.6 & -1 & -0.7 \\ 1.3 & 4.35 & -0.7 & -4 & -0.6 & -0.35 \\ -1.4 & -0.7 & 1.4 & 0 & 0 & 0.7 \\ -0.6 & -4 & 0 & 4 & 0.6 & 0 \\ -1 & -0.6 & 0 & 0.6 & 1 & 0 \\ -0.7 & -0.35 & 0.7 & 0 & 0 & 0.35 \end{bmatrix}$$

Element 4

This element is defined by the nodes 3, 4, and 5.

(a) The displacement-strain matrix $\{B\}$.

$$\{B\} = \frac{1}{72} \begin{bmatrix} 6 & 0 & -6 & 0 & 0 & 0 \\ 0 & 0 & 0 & -12 & 0 & 12 \\ 0 & 6 & -12 & -6 & 12 & 0 \end{bmatrix}$$

(b) The elasticity matrix $[D]$.

$$[D] = \frac{30{,}000}{0.91} \begin{bmatrix} 1 & 0.3 & 0 \\ 0.3 & 1 & 0 \\ 0 & 0 & 0.35 \end{bmatrix}$$

(c) The element stiffness matrix $[K]$.

$$[K]_4 = \frac{1}{72} \begin{bmatrix} 6 & 0 & 0 \\ 0 & 0 & 6 \\ -6 & 0 & -12 \\ 0 & -12 & -6 \\ 0 & 0 & 12 \\ 0 & 12 & 0 \end{bmatrix} \frac{30{,}000}{0.91} \begin{bmatrix} 1 & 0.3 & 0 \\ 0.3 & 1 & 0 \\ 0 & 0 & 0.35 \end{bmatrix}$$

$$\times \frac{1}{72} \begin{bmatrix} 6 & 0 & -6 & 0 & 0 & 0 \\ 0 & 0 & 0 & -12 & 0 & 12 \\ 0 & 6 & -12 & -6 & 12 & 0 \end{bmatrix} \times 1 \times 36$$

$$[K]_4 = 8250 \begin{bmatrix} 1 & 0 & -1 & -0.6 & 0 & 0.6 \\ 0 & 0.35 & -0.7 & -0.35 & 0.7 & 0 \\ -1 & -0.7 & 2.4 & 1.3 & -1.4 & -0.6 \\ -0.6 & -0.35 & 1.3 & 4.35 & -0.7 & -4 \\ 0 & 0.7 & -1.4 & -0.7 & 1.4 & 0 \\ 0.6 & 0 & -0.6 & -4 & 0 & 4 \end{bmatrix}$$

Because of the symmetry of the beam problem, the following equalities hold true.

$$[K]_5 = [K]_2$$
$$[K]_6 = [K]_1$$
$$[K]_7 = [K]_4$$
$$[K]_8 = [K]_3$$

6 The superposition of the element matrices to develop the total stiffness matrix of the structure is done by the scheme defined in expression (4-82). The known boundary conditions, which are $u_1 = v_1 = u_7 = u_8 = u_9 = 0$, will allow the simplification of the total 18 × 18 stiffness matrix into a 13 × 13 stiffness matrix as shown on pp. 132 and 133.

$[K]_{total} = (8250) \times$

	1	2	3	4	5	6	7	8	9	10	11	12	13	14	15	16	17	18
1	2.4	0	−1.4	0.7					0	−1.3	−1.0	0.6						
2	0	4.35	0.6	−4					−1.3	0	0.7	−0.35						
3	−1.4	0.6	4.8	0	−1.4	−0.6			−2	0								
4	0.7	−4	0	8.7	−0.7	−4			0	−0.7								
5			−1.4	−0.7	2.4	0	−1	−0.6	1.3	1.3								
6			−0.6	−4	0	4.35	−0.7	−0.35	−2.8	0								
7					−1	−0.7	4.8	0	−2.8	0			−1	0.7				
8					−0.6	−0.35	0	8.7	0	−8			0.6	−0.35				
9	0	−1.3	−2	0	1.3	−2.8	−2.8	0	9.6	0	−2.8	0	−1.3	−1.3	−2	0	0	1.3
10	−1.3	0	0	−0.7	1.3	0	0	−8	0	17.4	0	−8	0	0	0	−0.7	1.3	0
11	−1.0	0.7							−2.8	0	4.8	0					−1	−0.7
12	0.6	−0.35							0	−8	0	8.7					−0.6	−0.35
13							−1	0.6	−1.3	0			2.4	0	−1.4	0.7		
14							0.7	−0.35	−1.3	0			0	4.35	0.6	−4		
15									−2	0			−1.4	0.6	4.8	0	−1.4	−0.6
16									0	−0.7			0.7	−4	0	8.7	−0.7	−4
17									0	1.3	−1	−0.6			−1.4	−0.7	2.4	0
18									1.3	0	−0.7	−0.35			−0.6	−4	0	4.35

	1	2	3	4	5	6	7	8	9	10	11	12	13
1	4.8	0	-1.4	-0.6			-2						
2	0	8.7	-0.7	-4				-0.7					
3	-1.4	-0.7	2.4		-1	-0.6	1.3	1.3					
4	-0.6	-4		4.35	-0.7	-0.35	1.3						
5			-1	-0.7	4.8		-2.8				0.7		
6			-0.6	-0.35		8.7		-8			-0.35		
7	-2		1.3	1.3	-2.8		9.6		-2.8	0	-1.3	0	1.3
8		-0.7	1.3			-8		17.4	0	-8	0	-0.7	0
9							-2.8	0	4.8	0			-0.7
10							0	-8	0	8.7			-0.35
11					0.7	-0.35	-1.3	0			4.35	-4	
12							0	-0.7			-4	8.7	-4
13							1.3	0	-0.7	-0.35		-4	4.35

$[K]_{\text{total}} = 8250 \times$

$$
\begin{bmatrix}
5.357 & 0.5142 & 6.127 & 0.8011 & 3.485 & -3.724 & 2.250 & -4.310 & 0.4216 & -4.209 & -5.350 & -5.613 & -6.104 \\
0.5142 & 3.054 & 0.753 & 3.146 & 0.2192 & 1.992 & -0.2374 & 1.894 & 0.1466 & 1.821 & 1.769 & 1.865 & 1.956 \\
6.127 & 0.753 & 15.779 & 0.9208 & 7.565 & -10.204 & 3.393 & -11.670 & -0.3240 & -11.368 & -14.841 & -15.027 & -15.798 \\
0.8011 & 3.146 & 0.9208 & 6.257 & 0.4617 & 2.6218 & -0.6008 & 2.421 & 0.0013 & 2.323 & 2.003 & 2.225 & 2.412 \\
3.485 & 0.2192 & 7.565 & 0.4617 & 7.464 & -6.773 & 2.932 & -7.479 & 0.0451 & -7.337 & -10.828 & -10.831 & -11.418 \\
-3.728 & 1.992 & -10.204 & 2.6218 & -6.773 & 20.166 & -2.588 & 20.006 & 2.116 & 19.400 & 24.153 & 24.159 & 24.889 \\
2.250 & -0.2374 & 3.393 & -0.6008 & 2.932 & -2.588 & 3.202 & -2.909 & 1.235 & -2.849 & -3.062 & -3.630 & -4.325 \\
-4.310 & 1.894 & -11.670 & 2.421 & -7.479 & 20.006 & -2.909 & 21.412 & 2.143 & 20.752 & 25.417 & 25.528 & 26.357 \\
0.4216 & 0.1466 & -0.3240 & 0.0013 & 0.0451 & 2.116 & 1.235 & 2.143 & 3.843 & 2.134 & 4.168 & 3.954 & 4.056 \\
-4.209 & 1.821 & -11.368 & 2.323 & -7.337 & 19.400 & -2.849 & 20.752 & 2.134 & 21.527 & 25.018 & 25.152 & 26.054 \\
-5.350 & 1.769 & -14.841 & 2.003 & -10.828 & 24.153 & -3.062 & 25.417 & 4.168 & 25.018 & 42.835 & 40.536 & 40.874 \\
-5.613 & 1.865 & -15.027 & 2.225 & -10.831 & 24.159 & -3.630 & 25.528 & 3.954 & 25.152 & 40.536 & 41.253 & 41.680 \\
-6.104 & 1.956 & -15.798 & 2.412 & -11.418 & 24.889 & -4.325 & 26.357 & 4.056 & 26.054 & 40.874 & 41.680 & 45.158
\end{bmatrix}
\begin{bmatrix} u_2 \\ v_2 \\ u_3 \\ v_3 \\ u_4 \\ v_4 \\ u_5 \\ v_5 \\ u_6 \\ v_6 \\ u_7 \\ v_7 \\ u_8 \end{bmatrix}
=
\begin{bmatrix} 0 \\ 0 \\ 0 \\ -1 \\ 0 \\ -2 \\ 0 \\ 0 \\ 0 \\ -1 \\ 0 \\ 0 \\ 0 \end{bmatrix} \cdot 10^{-5}
$$

LONGHAND SOLUTION OF PLANE STRESS PROBLEMS

7. The calculations of the unknown displacements $\{\delta\}$ are performed by the general equation (4-72).

$$\{\delta\} = [K]^{-1}\{Q\}$$

The inversion of the total stiffness matrix $[K]$ is obtained by the aid of a digital computer (see p. 134).

$$\{\delta\} = \begin{bmatrix} 1.199 \\ -0.890 \\ 3.433 \\ -1.350 \\ 2.391 \\ -6.710 \\ 0.884 \\ -6.785 \\ -0.840 \\ -6.614 \\ -9.315 \\ -9.108 \\ -9.307 \end{bmatrix} \times 10^{-4} \text{ in.}$$

8. The computations of element stresses $\{\sigma\}$ are done according to the relations defined by Equations (4-36) and (4-39).

$$\{\sigma\} = [D]\{B\}\{\delta\}$$

The principal stresses are computed using the relations defined by expressions (3-10).

Element 1

$$\begin{Bmatrix} \sigma_x \\ \sigma_y \\ \tau_{xy} \end{Bmatrix} = 33{,}000 \begin{bmatrix} 1 & 0.3 & 0 \\ 0.3 & 1 & 0 \\ 0 & 0 & 0.35 \end{bmatrix} \frac{1}{72} \begin{bmatrix} 6 & 0 & 0 & 0 & -6 & 0 \\ 0 & 0 & 0 & -12 & 0 & 12 \\ 0 & 6 & -12 & 0 & 12 & -6 \end{bmatrix} \begin{bmatrix} 0 \\ 0 \\ 0.884 \times 10^{-4} \\ -6.785 \times 10^{-4} \\ -0.840 \times 10^{-4} \\ -6.614 \times 10^{-4} \end{bmatrix}$$

$$\begin{Bmatrix} \sigma_x \\ \sigma_y \\ \tau_{xy} \end{Bmatrix} = \begin{Bmatrix} -0.259 \\ -0.163 \\ -0.304 \end{Bmatrix} \text{ ksi}$$

The principal stresses are, according to expressions (3-10),

$$\sigma_{max} = -0.518 \text{ ksi}$$
$$\sigma_{min} = \sim 0.0 \text{ ksi}$$

Element 2

$$\begin{Bmatrix} \sigma_x \\ \sigma_y \\ \tau_{xy} \end{Bmatrix} = 33{,}000 \begin{bmatrix} 1 & 0.3 & 0 \\ 0.3 & 1 & 0 \\ 0 & 0 & 0.35 \end{bmatrix}$$

$$\times \frac{1}{72} \begin{bmatrix} 0 & 0 & 6 & 0 & -6 & 0 \\ 0 & 12 & 0 & -12 & 0 & 0 \\ 12 & 0 & -12 & 6 & 0 & -6 \end{bmatrix} \begin{bmatrix} 0 \\ 0 \\ 1.199 \times 10^{-4} \\ -0.890 \times 10^{-4} \\ 0.884 \times 10^{-4} \\ -6.785 \times 10^{-4} \end{bmatrix}$$

$$\begin{Bmatrix} \sigma_x \\ \sigma_y \\ \tau_{xy} \end{Bmatrix} = \begin{Bmatrix} -0.233 \\ -0.515 \\ -0.335 \end{Bmatrix} \text{ ksi}$$

Element 3

$$\begin{Bmatrix} \sigma_x \\ \sigma_y \\ \tau_{xy} \end{Bmatrix} = 33{,}000 \begin{bmatrix} 1 & 0.3 & 0 \\ 0.3 & 1 & 0 \\ 0 & 0 & 0.35 \end{bmatrix}$$

$$\times \frac{1}{72} \begin{bmatrix} 6 & 0 & 0 & 0 & -6 & 0 \\ 0 & 12 & 0 & -12 & 0 & 0 \\ 12 & 6 & -12 & 0 & 0 & -6 \end{bmatrix} \begin{bmatrix} 1.199 \times 10^{-4} \\ -0.890 \times 10^{-4} \\ 3.433 \times 10^{-4} \\ -1.350 \times 10^{-4} \\ 0.844 \times 10^{-4} \\ -6.785 \times 10^{-4} \end{bmatrix}$$

$$\begin{Bmatrix} \sigma_x \\ \sigma_y \\ \tau_{xy} \end{Bmatrix} = \begin{Bmatrix} -0.162 \\ -0.279 \\ -0.225 \end{Bmatrix} \text{ ksi}$$

Element 4

$$\begin{Bmatrix} \sigma_x \\ \sigma_y \\ \tau_{xy} \end{Bmatrix} = 33{,}000 \begin{bmatrix} 1 & 0.3 & 0 \\ 0.3 & 1 & 0 \\ 0 & 0 & 0.35 \end{bmatrix}$$

$$\times \frac{1}{72} \begin{bmatrix} 6 & 0 & -6 & 0 & 0 & 0 \\ 0 & 0 & 0 & -12 & 0 & 12 \\ 0 & 6 & -12 & -6 & 12 & 0 \end{bmatrix} \begin{bmatrix} 3.433 \times 10^{-4} \\ -1.350 \times 10^{-4} \\ 2.391 \times 10^{-4} \\ -6.710 \times 10^{-4} \\ 0.884 \times 10^{-4} \\ -6.785 \times 10^{-4} \end{bmatrix}$$

$$\begin{Bmatrix} \sigma_x \\ \sigma_y \\ \tau_{xy} \end{Bmatrix} = \begin{Bmatrix} -0.274 \\ -0.045 \\ -0.225 \end{Bmatrix} \text{ ksi}$$

Element 5

$$\begin{Bmatrix} \sigma_x \\ \sigma_y \\ \tau_{xy} \end{Bmatrix} = 33{,}000 \begin{bmatrix} 1 & 0.3 & 0 \\ 0.3 & 1 & 0 \\ 0 & 0 & 0.35 \end{bmatrix}$$

$$\times \frac{1}{72} \begin{bmatrix} 0 & 0 & 6 & 0 & -6 & 0 \\ 0 & 12 & 0 & -12 & 0 & 0 \\ 12 & 0 & -12 & 6 & 0 & -6 \end{bmatrix} \begin{bmatrix} 0.884 \times 10^{-4} \\ -6.785 \times 10^{-4} \\ 2.391 \times 10^{-4} \\ 6.710 \times 10^{-4} \\ 0 \\ -9.315 \times 10^{-4} \end{bmatrix}$$

$$\begin{Bmatrix} \sigma_x \\ \sigma_y \\ \tau_{xy} \end{Bmatrix} = \begin{Bmatrix} -0.645 \\ -0.156 \\ -0.039 \end{Bmatrix} \text{ ksi}$$

Element 6

$$\begin{Bmatrix} \sigma_x \\ \sigma_y \\ \tau_{xy} \end{Bmatrix} = 33{,}000 \times \begin{bmatrix} 1 & 0.3 & 0 \\ 0.3 & 1 & 0 \\ 0 & 0 & 0.35 \end{bmatrix}$$

$$\times \frac{1}{72} \begin{bmatrix} 6 & 0 & 0 & 0 & -6 & 0 \\ 0 & 0 & 0 & -12 & 0 & 12 \\ 0 & 6 & -12 & 0 & 12 & -6 \end{bmatrix} \begin{bmatrix} 0.884 \times 10^{-4} \\ -6.785 \times 10^{-4} \\ 0 \\ -9.315 \times 10^{-4} \\ 0 \\ -9.108 \times 10^{-4} \end{bmatrix}$$

$$\begin{Bmatrix} \sigma_x \\ \sigma_y \\ \tau_{xy} \end{Bmatrix} = \begin{Bmatrix} -0.277 \\ -0.186 \\ -0.223 \end{Bmatrix} \text{ ksi}$$

Element 7

$$\begin{Bmatrix} \sigma_x \\ \sigma_y \\ \tau_{xy} \end{Bmatrix} = 33{,}000 \begin{bmatrix} 1 & 0.3 & 0 \\ 0.3 & 1 & 0 \\ 0 & 0 & 0.35 \end{bmatrix}$$

$$\times \frac{1}{72} \begin{bmatrix} 6 & 0 & -6 & 0 & 0 & 0 \\ 0 & 0 & 0 & -12 & 0 & 12 \\ 0 & 6 & -12 & -6 & 12 & 0 \end{bmatrix} \begin{bmatrix} 0.884 \times 10^{-4} \\ -6.785 \times 10^{-4} \\ 0 \\ -9.180 \times 10^{-4} \\ 0 \\ -9.307 \times 10^{-4} \end{bmatrix}$$

$$\begin{Bmatrix} \sigma_x \\ \sigma_y \\ \tau_{xy} \end{Bmatrix} = \begin{Bmatrix} -0.210 \\ -0.036 \\ -0.223 \end{Bmatrix} \text{ ksi}$$

Element 8

$$\begin{Bmatrix} \sigma_x \\ \sigma_y \\ \tau_{xy} \end{Bmatrix} = 33{,}000 \begin{bmatrix} 1 & 0.3 & 0 \\ 0.3 & 1 & 0 \\ 0 & 0 & 0.35 \end{bmatrix}$$

$$\times \frac{1}{72} \begin{bmatrix} 6 & 0 & 0 & 0 & -6 & 0 \\ 0 & 12 & 0 & -12 & 0 & 0 \\ 12 & 6 & -12 & 0 & 0 & -6 \end{bmatrix} \begin{bmatrix} -0.840 \times 10^{-4} \\ -6.614 \times 10^{-4} \\ 0.844 \times 10^{-4} \\ -6.785 \times 10^{-4} \\ 0 \\ -9.037 \times 10^{-4} \end{bmatrix}$$

$$\begin{Bmatrix} \sigma_x \\ \sigma_y \\ \tau_{xy} \end{Bmatrix} = \begin{Bmatrix} 0.203 \\ -0.025 \\ -0.072 \end{Bmatrix} \text{ ksi}$$

This ends the longhand analysis of the simply supported beam by the finite-element method.

PROBLEMS

5-1 Perform a longhand finite element analysis of a thin square plate fixed at one end and pulled by a uniform load at the opposite end. Use constant-strain triangular elements to idealize the structure. The modulus of elasticity is 30,000 ksi and Poisson ratio is 0.3. The geometry and loading intensity are given in Fig. 5-11. The thickness of the plate is 0.5 in.

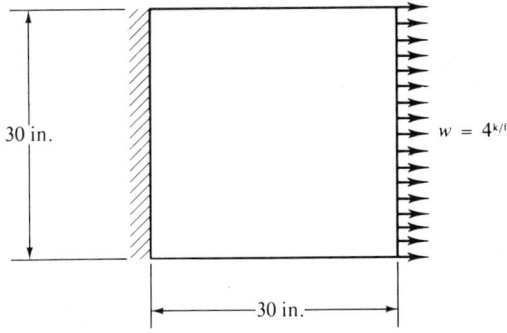

Figure 5-11

5-2 Perform the longhand finite element analysis of the given beam. The modulus of elasticity is 29,000 ksi and Poisson ratio is 0.3. The load is uniform of intensity 5 kips/linear ft. Use a coarse mesh for the simplicity of the analysis. The beam is shown in Fig. 5-12. Use the symmetry of the structure for the formulation of your analysis.

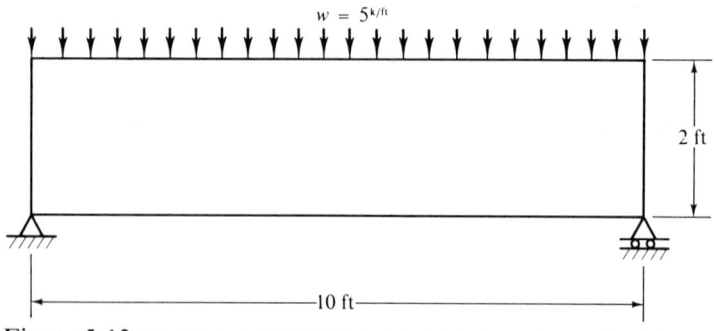

Figure 5-12

Analysis of Axisymmetric Solids

INTRODUCTION

The application of the finite element method to two-dimensional stress analysis problems, using constant-strain triangular elements, has been presented in Chapter 4. The basic approach of this method idealized the given structural system by various finite elements joined together at defined nodal points. This is an effective geometric approximation of a continuous media. It is this geometric approximation that allows the generation of governing equilibrium equations which are linear, hence mathematically simple to solve. If a continuous media is considered, its governing differential equations will present various difficulties depending on the complexity of the problem.

Once the element types are selected for the idealization, general displacement functions are assumed to describe the displacement behavior of the elements. These functions should define correctly the element interior as well as the boundary deflections, otherwise the compatibility conditions will not be satisfied and consequently the scientific value of the finite element method will be lost. The proper differentiation of these displacement functions will yield the element strain relations. In plane stress problems, the assumed functions being linear, the strains will have constant values. However, in other cases the strains can be functions of variables. Since the strains are related to the stresses by well-defined elasticity matrices, it is possible to determine the stress values in elements. The preceding paragraphs which concisely describe the logic used by the general finite element method is applicable to all analysis problems, simple or complex.

In structural analysis, axisymmetric systems do occur often and need to be considered. A spherical solid is an axisymmetric structure which has the z axis as its axis of symmetry. Figure 6-1 shows a vertical section

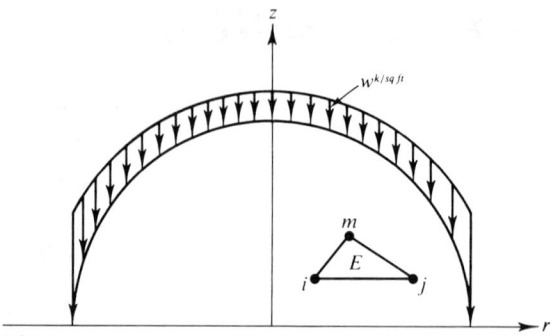

Figure 6-1

of the sphere cut by a plane passing through the origin. The sphere is uniformly loaded by gravity forces. Since it is geometrically and loading-wise symmetric about the z axis, each internal point will have two degrees of freedom for displacements. They will be in radial (r) and in vertical (z) directions. If one relates the r-z coordinates to the X-Y coordinates of the plane stress analysis discussed in Chapter 4, it will be obvious that axisymmetric solids can be treated in a similar manner to two-dimensional plane stress problems. An element triangular in cross section, ijm, as shown in Fig. 6-1, can be used to develop a mesh for the structure.

AXISYMMETRIC VERSUS PLANE STRESS ANALYSIS

The plane stress analysis used elements which were lying on a plane field, and their interconnections at nodal points occurred in the same planar surface. In axisymmetric analysis the elements lie in three-dimensional space. The elements used are triangular tori (sing., torus) whose any random vertical cross section will be a plane triangle.

The nodal points related to elements of plane stress analysis problems will be actual points in a plane. The nodal points of an axisymmetric triangular element will be circumferential lines. Hence the mathematical procedures and operations required to develop the element stiffness matrix will be more complex. An integration operation will be necessary to satisfy the calculation along circumferential boundary lines. A triangular torus element associated with the analysis of axisymmetric structural systems is given in Fig. 6-2.

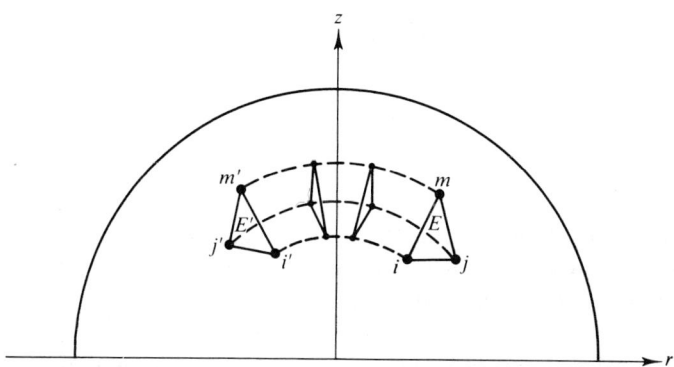

Figure 6-2

In the plane stress elasticity problem, stresses existed only in the X and Y directions.

In the axisymmetric analysis problems, the radial displacements will develop circumferential strains which will induce stresses in the same directions. Therefore in the mathematical formulation of the problem, the circumferential strains and stresses must be considered.

Triangular torus elements are preferred to idealize the axisymmetric system since they are capable of simulating complex surfaces and also they are comparatively simple to work with.

It is noted that the basic steps of the finite element method which are applied to the analysis of plane stress problems, are also suitable for the analysis of axisymmetric structures. The only major variations will be in the generation of element stiffness matrix and load matrix.

AXISYMMETRIC ANALYSIS

A triangular axisymmetric element will have three nodes and each node will have two degrees of freedom. The radial displacement is expressed by u and the vertical displacement by v, to be compatible with the displacement assumptions of the plane stress analysis. The nodes and related

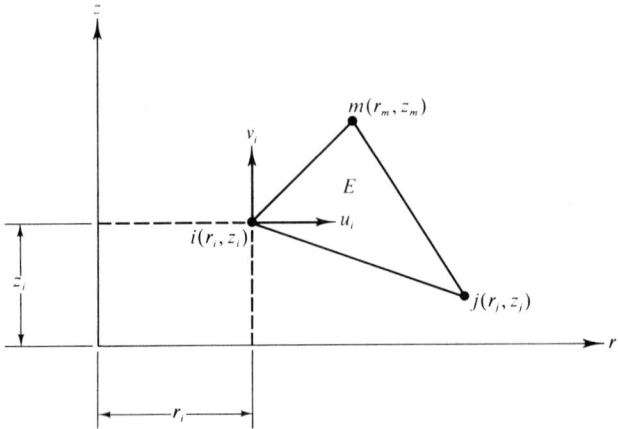

Figure 6-3

degrees of freedom are shown in Fig. 6-3. The theoretical development will follow the steps presented in Chapter 4 which dealt with the theory applicable to the analysis of plane stress problems using constant strain triangular elements.

a. The unknowns of the problem are the nodal displacements, $\{\delta\}$. Each node will have two degrees of freedom, one in the r direction and the other in the z direction. A triangular axisymmetric element will present six degrees of freedom.

$$\{\delta\} = \begin{Bmatrix} \delta_i \\ \delta_j \\ \delta_m \end{Bmatrix} \quad \begin{Bmatrix} u_i \\ v_i \\ u_j \\ v_j \\ u_m \\ v_m \end{Bmatrix} \tag{6-1}$$

b. The general displacement functions to describe the behavior of elements will be similar to expressions (4-2).

$$\{\phi\} = \begin{Bmatrix} u(r,z) \\ v(r,z) \end{Bmatrix} = \begin{Bmatrix} a_1 + a_2 r + a_3 z \\ a_4 + a_5 r + a_6 z \end{Bmatrix} \tag{6-2}$$

where $\{\phi\}$ is the general displacement function; $a_1, a_2, a_3, a_4, a_5, a_6$ are the undetermined coefficients of the displacement function. In matrix notation, $\{\phi\}$ is written as shown by expression (6-3).

$$\{\phi\} = \begin{bmatrix} 1 & r & z & 0 & 0 & 0 \\ 0 & 0 & 0 & 1 & r & z \end{bmatrix} \begin{Bmatrix} a_1 \\ a_2 \\ a_3 \\ a_4 \\ a_5 \\ a_6 \end{Bmatrix} \qquad (6\text{-}3)$$

The values of the unknown coefficients of the displacement functions are obtained by evaluating the displacements functions at the nodal points. This procedure will introduce six linear equations.

$$\begin{aligned}
u_i &= a_1 + a_2 r_i + a_3 z_i \\
u_j &= a_1 + a_2 r_j + a_3 z_j \\
u_m &= a_1 + a_2 r_m + a_3 z_m \\
v_i &= a_4 + a_5 r_i + a_6 z_i \\
v_j &= a_4 + a_5 r_j + a_6 z_j \\
v_m &= a_4 + a_5 r_m + a_6 z_m
\end{aligned} \qquad (6\text{-}4)$$

Operating on the system of linear equations (6-4), the coefficients $a_1, a_2, \ldots, a_6$ are written as functions of nodal displacements:

$$\begin{Bmatrix} a_1 \\ a_2 \\ a_3 \end{Bmatrix} = \begin{bmatrix} 1 & r_i & z_i \\ 1 & r_j & z_j \\ 1 & r_m & z_m \end{bmatrix}^{-1} \begin{Bmatrix} u_i \\ u_j \\ u_m \end{Bmatrix} \qquad (6\text{-}5)$$

$$\begin{Bmatrix} a_4 \\ a_5 \\ a_6 \end{Bmatrix} = \begin{bmatrix} 1 & r_i & z_i \\ 1 & r_j & z_j \\ 1 & r_m & z_m \end{bmatrix}^{-1} \begin{Bmatrix} v_i \\ v_j \\ v_m \end{Bmatrix} \qquad (6\text{-}6)$$

Performing the inverse operations in Equations (6-5) and (6-6), new expressions are derived:

$$\begin{Bmatrix} a_1 \\ a_2 \\ a_3 \end{Bmatrix} = \frac{1}{2A} \begin{bmatrix} \alpha_i & \alpha_j & \alpha_m \\ \beta_i & \beta_j & \beta_m \\ \gamma_i & \gamma_j & \gamma_m \end{bmatrix} \begin{Bmatrix} u_i \\ u_j \\ u_m \end{Bmatrix} \qquad (6\text{-}7)$$

$$\left\{\begin{matrix}a_4\\a_5\\a_6\end{matrix}\right\} = \frac{1}{2A}\begin{bmatrix}\alpha_i & \alpha_j & \alpha_m\\ \beta_i & \beta_j & \beta_m\\ \gamma_i & \gamma_j & \gamma_m\end{bmatrix}\left\{\begin{matrix}v_i\\v_j\\v_m\end{matrix}\right\} \qquad (6\text{-}8)$$

where

$$\begin{aligned}\alpha_i &= r_j z_m - z_j r_m\\ \alpha_j &= r_m z_i - z_m r_i\\ \alpha_m &= r_i z_j - z_i r_j\\ \beta_i &= z_j - z_m\\ \beta_j &= z_m - z_i\\ \beta_m &= z_i - z_j\\ \gamma_i &= r_m - r_j\\ \gamma_j &= r_i - r_m\\ \gamma_m &= r_j - r_i\end{aligned} \qquad (6\text{-}9)$$

It is noted that the above expressions can be obtained by proper permutations when one of α, β, and γ expressions is defined.

To develop a general expression for the displacement function $\{\phi\}$, let us define three new expressions which are described by Equation (4-21).

$$\begin{aligned}N_i &= (\alpha_i + \beta_i r + \gamma_i z)\\ N_j &= (\alpha_j + \beta_j r + \gamma_j z)\\ N_m &= (\alpha_m + \beta_m r + \gamma_m z)\end{aligned} \qquad (6\text{-}10)$$

Then the displacement function is given below according to Equations (4-22), (4-23), and (4-24).

$$\{\phi\} = \frac{1}{2A}\begin{Bmatrix}N_i & 0 & N_j & 0 & N_m & 0\\ 0 & N_i & 0 & N_j & 0 & N_m\end{Bmatrix}\left\{\begin{matrix}u_i\\v_i\\u_j\\v_j\\u_m\\v_m\end{matrix}\right\} \qquad (6\text{-}11)$$

which can be summarized in generalized matrix notation as

$$\{\phi\} = \frac{1}{2A} \cdot \{N\} \cdot \{\delta\} \qquad (6\text{-}12)$$

The unknown nodal displacements are computed using the principle of stationary potential energy as given by Equation (4-72).

c. The element strains $\{\epsilon\}$ are obtained by proper differentiation of the displacement polynomials and also as a direct relation of radial displacement. It is noted that a fourth strain component, circumferential strain (ϵ_θ), is considered in the axisymmetric analysis as discussed in the first part of this chapter. Expression (6-13) defines all the strain components:

$$\{\epsilon\} = \begin{Bmatrix} \epsilon_r \\ \epsilon_z \\ \epsilon_\theta \\ \gamma_{rz} \end{Bmatrix} = \begin{Bmatrix} \dfrac{\partial u}{\partial r} \\ \dfrac{\partial v}{\partial z} \\ \dfrac{u}{r} \\ \dfrac{\partial u}{\partial z} + \dfrac{\partial v}{\partial r} \end{Bmatrix} \qquad (6\text{-}13)$$

According to the discussions presented in Chapter 4 which converge to expression (4-35), we can write a significant matrix expression for the strain matrix $\{\epsilon\}$ as shown by Equation (6-14).

$$\{\epsilon\} = \begin{Bmatrix} \epsilon_r \\ \epsilon_z \\ \epsilon_\theta \\ \gamma_{ry} \end{Bmatrix} = \frac{1}{2A} \cdot \{B_i B_j B_m\} \begin{Bmatrix} \delta_i \\ \delta_j \\ \delta_m \end{Bmatrix} \qquad (6\text{-}14)$$

Using Equations (6-2) and (6-13), one can also write the following relation:

$$\begin{Bmatrix} \epsilon_r \\ \epsilon_z \\ \epsilon_\theta \\ \gamma_{ry} \end{Bmatrix} = \begin{Bmatrix} a_2 \\ a_6 \\ \dfrac{a_1}{r} + a_2 + \dfrac{a_3 z}{r} \\ a_3 + a_5 \end{Bmatrix} \qquad (6\text{-}15)$$

In matrix expanded notation, Equation (6-15) will have the form given by expression (6-16).

148 ANALYSIS OF AXISYMMETRIC SOLIDS

$$\begin{Bmatrix} \epsilon_r \\ \epsilon_z \\ \epsilon_\theta \\ \gamma_{ry} \end{Bmatrix} = \begin{Bmatrix} 0 & 1 & 0 & 0 & 0 & 0 \\ 0 & 0 & 0 & 0 & 0 & 1 \\ \dfrac{1}{r} & 1 & \dfrac{z}{r} & 0 & 0 & 0 \\ 0 & 0 & 1 & 0 & 1 & 0 \end{Bmatrix} \begin{Bmatrix} a_1 \\ a_2 \\ a_3 \\ a_4 \\ a_5 \\ a_6 \end{Bmatrix} \qquad (6\text{-}16)$$

Substituting Equations (4-13) and (4-14) in Equation (6-16), we obtain expression (6-17), which relates the strains to the nodal displacements

$$\begin{Bmatrix} \epsilon_r \\ \epsilon_z \\ \epsilon_\theta \\ \gamma_{ry} \end{Bmatrix} = \frac{1}{2A}$$

$$\cdot \begin{Bmatrix} \beta_i & 0 & \beta_j & 0 & \beta_m & 0 \\ 0 & \gamma_i & 0 & \gamma_j & 0 & \gamma_m \\ \dfrac{\alpha_i}{r} + \beta_i + \dfrac{\gamma_i z}{r} & 0 & \dfrac{\alpha_j}{r} + \beta_j + \dfrac{\gamma_j z}{z} & 0 & \dfrac{\alpha_m}{r} + \beta_m + \dfrac{\gamma_m z}{r} & 0 \\ \gamma_i & \beta_i & \gamma_j & \beta_j & \gamma_m & \beta_m \end{Bmatrix} \begin{Bmatrix} u_i \\ v_i \\ u_j \\ v_j \\ u_m \\ v_m \end{Bmatrix}$$

$$(6\text{-}17)$$

Expression (6-17) can be generalized by the following equation as given by Equation (4-36):

$$\{\epsilon\} = \{B\}\{\delta\} \qquad (6\text{-}18)$$

We do note that the matrix $\{B\}$ of Equation (6-18) includes the coordinates r and z. Therefore the strains will not be constant as was the case in the plane stress analysis of two-dimensional problems. The only case where the strains in the axisymmetric analysis will be constant is when $\epsilon_\theta = u/r$ is constant. This can be possible if u and r are proportional to each other.

Equation (6-18) will produce the element strain values once the $\{\delta\}$ matrix is determined according to Equation (4-72).

d. The element stresses are related to the element strains by the elasticity matrix $[D]$.

$$\{\sigma\} = [D]\{\epsilon\} \qquad (6\text{-}19)$$

For an element made of isotropic material, the strains are related to the stresses by the following elasticity equation:

$$\begin{aligned}
\epsilon_r &= -\nu\frac{\sigma_z}{E} + \frac{\sigma_r}{E} - \nu\frac{\sigma_\theta}{E} \\
\epsilon_z &= \frac{\sigma_z}{E} - \nu\frac{\sigma_r}{E} - \nu\frac{\sigma_\theta}{E} \\
\epsilon_\theta &= -\nu\frac{\sigma_z}{E} - \nu\frac{\sigma_r}{E} + \frac{\sigma_\theta}{E} \\
\gamma_{ry} &= \frac{2(1+\nu)}{E}\tau_{ry}
\end{aligned} \qquad (6\text{-}20)$$

Expression (6-20) in matrix notation will be written as follows:

$$\begin{Bmatrix}\epsilon_r \\ \epsilon_z \\ \epsilon_\theta \\ \gamma_{ry}\end{Bmatrix} = \frac{1}{E}\begin{bmatrix} 1 & -\nu & -\nu & 0 \\ -\nu & 1 & -\nu & 0 \\ -\nu & -\nu & 1 & 0 \\ 0 & 0 & 0 & 2(1+\nu) \end{bmatrix}\begin{Bmatrix}\sigma_r \\ \sigma_z \\ \sigma_\theta \\ \tau_{rz}\end{Bmatrix} \qquad (6\text{-}21)$$

The matrix equation (6-21) is rearranged for the direct solution of the element stress values:

$$\begin{Bmatrix}\sigma_r \\ \sigma_z \\ \sigma_\theta \\ \tau_{ry}\end{Bmatrix} = E\begin{bmatrix} 1 & -\nu & -\nu & 0 \\ -\nu & 1 & -\nu & 0 \\ -\nu & -\nu & 1 & 0 \\ 0 & 0 & 0 & 2(1+\nu) \end{bmatrix}^{-1}\begin{Bmatrix}\epsilon_r \\ \epsilon_z \\ \epsilon_\theta \\ \gamma_{rz}\end{Bmatrix} \qquad (6\text{-}22)$$

After performing the inverse operation, using a logic described by Equations (4-53), (4-54), and (4-55), expression (6-22) is rewritten:

$$\begin{Bmatrix}\sigma_r \\ \sigma_z \\ \sigma_\theta \\ \tau_{ry}\end{Bmatrix} = \frac{E}{(1+\nu)(1-2\nu)}\begin{bmatrix} \nu & (1-\nu) & \nu & 0 \\ (1-\nu) & \nu & \nu & 0 \\ \nu & \nu & (1-\nu) & 0 \\ 0 & 0 & 0 & \frac{1-2\nu}{2} \end{bmatrix}\begin{Bmatrix}\epsilon_r \\ \epsilon_z \\ \epsilon_\theta \\ \gamma_{rz}\end{Bmatrix} \qquad (6\text{-}23)$$

Expression (6-23) will yield the element stress values as a function of element strain.

e. All the previously developed expressions depend on the values of the nodal displacements. Once they are numerically computed, then the element strains and element stresses can be numerically computed according to Equations (6-18) and (6-19).

The determination of the nodal displacements will require the generation of a structural stiffness matrix $[K]$, which relates the applied loads to nodal displacements as expressed by Equation (4-72)

$$\{\delta\} = [K]^{-1}\{Q\}$$

The total stiffness matrix $[K]$ is generated by proper superposition of element stiffness matrices as defined by the summation expression (4-69):

$$[K] = \sum_1^n [k]$$

The development of element stiffness matrix can be summarized by the following steps.

1. Since the finite element method is based on the principle of stationary total potential energy, this quantity needs to be obtained. The strain energy stored in the system is defined by the product of the strain with the resulting stresses. The external energy is defined by the product of nodal displacements with the corresponding nodal forces. The algebraic sum of these energies yields the total potential energy of the structural system.

2. According to the stationary nature of the total potential energy, its variation is zero as shown by Equation (3-15)

$$\delta V = 0$$

where V defines the total potential energy. The proper differentiation of the total potential energy expression V with respect to the displacement components will generate a system of linear equations which correlate nodal forces to nodal displacements, Equation (4-71). The coefficient matrix which relates the nodal forces to the nodal displacements is the element stiffness matrix. The equations given by expression (4-71) define the equilibrium conditions at nodal points.

Therefore, expression (4-69) is used to obtain the total structural stiffness matrix $[K]$, which will allow the computations of all nodal point displacements by Equation (4-72). The element stiffness matrix is numerically computed according to the general expressions (4-68) and (4-77). The integral expression (4-68)

$$[k] = \int_V \{B\}^T[D]\{B\}\, dV$$

needs to be integrated along the circumferential boundary. It can be rewritten as

$$[k] = 2\pi \int_{area} r \cdot \{B\}^T[D]\{B\} \cdot dr\, dz \qquad (6\text{-}24)$$

The matrix quantities $\{B\}$ and $[D]$ are defined respectively by expressions (6-17) and (6-23). Since the $\{B\}$ matrices do depend on the coordinates, the integration operations will be more complex than the ones encountered in plane stress problems. There are two general approaches to evaluate the integrals.

(a) The matrix expansion of the expression is obtained by multiplication of the three matrix quantities and then integrating all the terms individually.

(b) The evaluation of the $\{B\}$ matrices is found at the centroids of the elements. The coordinates of the centroidal points are expressed as functions of the coordinates of nodal points, and these centroidal coordinates are used in $\{B\}$ matrices. This procedure will produce acceptable results. If more accurate results are desired, then smaller elements should be used.

There are other more elaborate mathematical routines which might be used if a sophisticated solution approach is desired.

f. The externally applied loads need to be expressed in a form to be compatible with axisymmetric problem. If Q_{unit} represents the radial component of load per linear foot of the circumferential boundary of the element, the total load to be considered in the analysis will be obtained by integrating the Q_{unit} component along the boundary of the element. The total load Q can be expressed as follows.

$$Q = 2\pi r Q_{unit} \qquad (6\text{-}25)$$

In the vertical direction, the load to be considered will be

$$Z = 2\pi r Z_{unit} \qquad (6\text{-}26)$$

where Z defines the vertical element load. Other load components can be computed accordingly if they exist. The possible forces, such as body forces and thermal forces, were discussed in Chapter 4. One can conclude that the solution of the analysis of axisymmetric structures can be obtained without much difficulty once the total structural stiffness matrix and load matrices are generated. The rest of the operations will follow the general finite element approach redefined below.

1 Numerical computations of nodal displacements, Equation (4-71):
$$\{\delta\} = [K]^{-1}\{Q\}$$
2 Numerical computation of element strain, Equation (4-36):
$$\{\epsilon\} = \{B\}\{\delta\}$$
3 Numerical computation of element strain, Equation (4-39):
$$\{\sigma\} = [D]\{\epsilon\}$$

It might be beneficial to discuss the value of axisymmetric analysis with respect to the use of the finite element method.

EVALUATION OF AXISYMMETRIC ANALYSIS

Since the finite-element analysis of axisymmetric elastic systems follow the same logical steps as the analysis of two-dimensional plane stress problems, its applications have been quick and effective. There are many computer programs to analyze two-dimensional plane stress-plane strain problems. These programs can easily be adapted to the analysis of axisymmetric structural systems. The only changes required in these programs are

1 Rearrangement of the logic of the generation of element stiffness matrices. Since the elements have circumferential boundaries, an integration operation along these boundaries needs to be performed.
2 Development of a new scheme to generate the load matrices due to the same reasons mentioned in Item 1.

Therefore, the direct applications of the finite element method to axisymmetric problems have been rather quick.

There are many structural systems which are classified as axisymmetric: solid or hollow spherical structures; thin- or thick-walled cylindrical systems; nuclear reactor vessels; stress distribution problems in continuous media, and rocket nozzles. All of these structures have great practical importance in the profession of structural engineering. They all need solutions to the problems of efficient and optimum design. The finite element method using axisymmetric structural analysis concepts can develop solutions to all of these problems. A triangular torus element is used in the presentation of this chapter due to its simplicity and ability to describe geometric boundaries with ease. However, there are other elements such as rectangles and other quadrilaterals which are also adaptable to the axisymmetric analysis. The engineer will decide on the selection of the type of elements in practical problems.

REFERENCES

1   R. W. CLOUGH, "The Finite Element Method in Plane Stress Analysis," *Proceedings of the American Society of Civil Engineers,* 345–378 (1960).
2   S. TIMOSHENKO, and J. H. GOODIER, *Theory of Elasticity.* New York: McGraw-Hill, 1970.
3   C. K. WANG, *Applied Elasticity.* New York: McGraw-Hill, 1953.
4   R. W. CLOUGH and Y. RASHID, "Finite Element Analysis of Axisymmetric Solids," *Proceedings of the American Society of Civil Engineers,* EM 1, 71–85 (February 1965).
5   Y. RASHID, "Analysis of Axisymmetric Composite Structures by the Finite Element Method," *Nuclear Engineering and Design* **3,** 163–182 (1966).
6   E. L. WILSON, "Structural Analysis of Axisymmetric Solids," *AIAA Journal* **3,** 12, 2269–2274 (December 1965).

Additional Element Types for Finite Element Analysis 7

INTRODUCTION

Finite elements are used for the idealization of continuous media for the purpose of allowing their analysis by matrix methods. In the discussions of the previous chapters, constant-strain triangular elements were used. The reason for the choice of that particular element was its simplicity for the presentation of the basic concepts and logic of the finite element theory. Since the reader is now familiar with the background of the method, other finite element types will now be introduced.

Any geometric shapes can be used to generate a mesh for the structural system to be analyzed. However, mathematical and practical constraints limit the possible varieties of elements. The mathematical requirements have to be fully obeyed in order to secure the acceptability of the results of the analysis. These requirements are that the assumed displacement function for the element must satisfy both the displacement behavior of the element and the displacement continuities across the element boundaries. The approximate solutions obtained by the finite element

156 ADDITIONAL ELEMENT TYPES FOR FINITE ELEMENT ANALYSIS

method should converge to the true solutions as the element sizes diminish. It is understood that as the elements become smaller, their strain expressions tend to have a constant value. The assumed displacement functions should mathematically reflect this fact.

From a practical point of view, it is not wise to select unorthodox geometric shapes as finite elements, since they might either be incapable of properly idealizing the continuous surface or be hard to define mathematically. Because of these practical and mathematical constraints, the number of available and efficient elements are limited. They fit into the finite-element method just like the constant-strain triangular element. However, their geometric characteristics change the matrix expressions which are used in the derivation of the method. As the element becomes more complex, the mathematical operations associated with their behavior become more complex also. For example, a constant-strain triangular element had three nodes and six degrees of freedom; if a rectangular plane stress element is involved, there are four nodes and eight degrees of freedom to be considered. If a linear-strain triangular element is introduced, then one has to use six nodes and twelve displacements. Thus, as the elements become refined or different the associated mathematical operations vary to some extent.

However, there are advantages of using refined elements, since they fit some problems better and yield more accurate results even when a coarse mesh of idealization is used. There are also complex problems which require sophisticated elements. The reader should be familiar with the characteristics of a few other finite elements in order to use the method properly and efficiently.

In this chapter several new elements will be presented and discussed. Since the majority of steps in the discussions will be parallel and similar to those presented in Chapter 4, there will be references to that chapter to avoid excessive repetition.

1. RECTANGULAR PLANE STRESS ELEMENT

A rectangular plane stress element has four nodal points and each node has two degrees of displacement freedom, one in the X direction and the other in the Y direction. The element's total number of degrees of freedom will be eight. A typical rectangular element is shown in Fig. 7-1. Since the rectangular element has four nodes, it will permit the assumption of a displacement polynomial with four constants. These constants are determined by the solution of the system of equations obtained by

1. RECTANGULAR PLANE STRESS ELEMENT

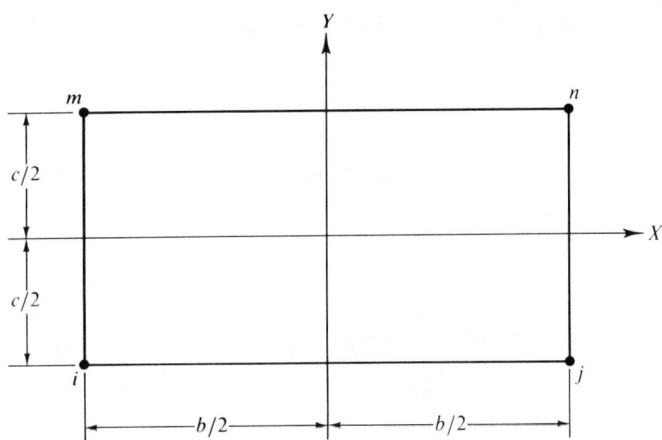

Figure 7-1

substituting the known coordinates of the four nodal points in the assumed displacement function [1].

There are several forms of a quadratic function. The one to be selected has to approximate the displacement behavior of the element and also satisfy the displacement compatibility among the elements. To secure this compatibility, the displacement function needs to be linear in nature along the boundaries of adjoining rectangular elements. The assumed displacement polynomial must have a linear form at the boundaries of the element. This restriction will limit the choice of acceptable quadratic functions. The following expressions will satisfy the constraints of the problem, since along the edges one of the variables will have a constant value making these expressions linear.

$$u(x,y) = a_1 + a_2 x + a_3 y + a_4 xy$$
$$v(x,y) = a_5 + a_6 x + a_7 y + a_8 xy \qquad (7\text{-}1)$$

where $u(x,y)$ is the X displacements, $v(x,y)$ the Y displacements of the element points, and $a_1, a_2, \ldots, a_8$ the coefficients to be determined.

When expression (7-1) is compared with expression (4-2), we observe that the rectangular plane stress element introduces two additional coefficients to be determined. The incorporation of the rectangular element in the plane stress finite element analysis will follow the same steps used in Chapter 4 which treated the constant-strain triangular element. The general development of the theory will be concisely rediscussed for the benefit of the reader and for added clarity of presentation.

a. The nodal displacements are the unknowns of the analysis. The rectangular element has four nodes and each node has two degrees of

158 ADDITIONAL ELEMENT TYPES FOR FINITE ELEMENT ANALYSIS

freedom. The displacement matrix $\{\delta\}$ will be similar to the one given by Equation (4-1).

$$\{\delta\} = \begin{Bmatrix} \delta_i \\ \delta_j \\ \delta_m \\ \delta_n \end{Bmatrix} = \begin{Bmatrix} u_i \\ v_i \\ u_j \\ v_j \\ u_m \\ v_m \\ u_n \\ v_n \end{Bmatrix} \qquad (7\text{-}2)$$

b. The general displacement function is given by Equation (7-1) which satisfies the mathematical requirements of the method as discussed in the previous section of this chapter.

$$\{\phi\} = \begin{Bmatrix} u(x,y) \\ v(x,y) \end{Bmatrix} = \begin{Bmatrix} a_1 + a_2 x + a_3 y + a_4 xy \\ a_5 + a_6 x + a_7 y + a_8 xy \end{Bmatrix} \qquad (7\text{-}3)$$

The matrix $\{\phi\}$ is the general displacement matrix. Expression (7-3) can be expanded in the form

$$\{\phi\} = \begin{bmatrix} 1 & x & y & xy & 0 & 0 & 0 & 0 \\ 0 & 0 & 0 & 0 & 1 & x & y & xy \end{bmatrix} \begin{Bmatrix} a_1 \\ a_2 \\ a_3 \\ a_4 \\ a_5 \\ a_6 \\ a_7 \\ a_8 \end{Bmatrix} \qquad (7\text{-}4)$$

To determine the constants $a_1, a_2, \ldots, a_8$, the coordinates of the four nodes are substituted in the general displacement function as done in Equations (4-4) and (4-5). This operation will produce a system of equations shown by Equations (7-5) and (7-6).

1. RECTANGULAR PLANE STRESS ELEMENT

$$\begin{aligned} u_i &= a_1 + a_2 x_i + a_3 y_i + a_4 x_i y_i \\ u_j &= a_1 + a_2 x_j + a_3 y_j + a_4 x_j y_j \\ u_m &= a_1 + a_2 x_m + a_3 y_m + a_4 x_m y_m \\ u_n &= a_1 + a_2 x_n + a_3 y_n + a_4 x_n y_n \end{aligned} \quad (7\text{-}5)$$

and

$$\begin{aligned} v_i &= a_5 + a_6 x_i + a_7 y_i + a_8 x_i y_i \\ v_j &= a_5 + a_6 x_j + a_7 y_j + a_8 x_j y_j \\ v_m &= a_5 + a_6 x_m + a_7 y_m + a_8 x_m y_m \\ v_n &= a_5 + a_6 x_n + a_7 y_n + a_8 x_n y_n \end{aligned} \quad (7\text{-}6)$$

Since the operations necessary to solve the system given by Equations (7-5) and (7-6) are similar, only Equation (7-5) will be solved showing the major steps. In matrix notation, Equation (7-5) is written as follows:

$$\begin{Bmatrix} u_i \\ u_j \\ u_m \\ u_n \end{Bmatrix} = \begin{bmatrix} 1 & x_i & y_i & x_i y_i \\ 1 & x_j & y_j & x_j y_j \\ 1 & x_m & y_m & x_m y_m \\ 1 & x_n & y_n & x_n y_n \end{bmatrix} \begin{Bmatrix} a_1 \\ a_2 \\ a_3 \\ a_4 \end{Bmatrix} \quad (7\text{-}7)$$

Since the desired values are the coefficients, matrix equation (7-7) is rearranged into the new form

$$\begin{Bmatrix} a_1 \\ a_2 \\ a_3 \\ a_4 \end{Bmatrix} = \begin{bmatrix} 1 & x_i & y_i & x_i y_i \\ 1 & x_j & y_j & x_j y_j \\ 1 & x_m & y_m & x_m y_m \\ 1 & x_n & y_n & x_n y_n \end{bmatrix}^{-1} \begin{Bmatrix} u_i \\ u_j \\ u_m \\ u_n \end{Bmatrix} \quad (7\text{-}8)$$

If the numerical coordinates of the nodes which are shown in Fig. 7-2 are substituted in expression (7-8), and the inversion operation is performed, one obtains the values of coefficients as functions of unknown nodal displacements.

$$\begin{Bmatrix} a_1 \\ a_2 \\ a_3 \\ a_4 \end{Bmatrix} = \frac{1}{4} \begin{bmatrix} 1 & 1 & 1 & 1 \\ -\frac{1}{b} & \frac{1}{b} & \frac{1}{b} & -\frac{1}{b} \\ -\frac{1}{c} & -\frac{1}{c} & \frac{1}{c} & \frac{1}{c} \\ \frac{1}{bc} & -\frac{1}{bc} & \frac{1}{bc} & -\frac{1}{bc} \end{bmatrix} \begin{Bmatrix} u_i \\ u_j \\ u_m \\ u_n \end{Bmatrix} \quad (7\text{-}9)$$

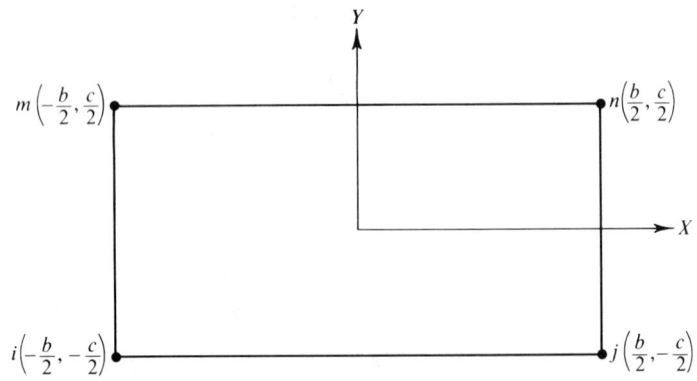
Figure 7-2

If the values of the coefficients given by Equation (7-9) are placed in expression (7-3), one obtains the new form of the general displacement function given by expression (7-10).

$$u(x, y) = a_1 + a_2 x + a_3 y + a_4 xy$$
$$v(x, y) = a_5 + a_6 x + a_7 y + a_8 xy$$

Since

$$a_1 = \frac{1}{4}(u_i + u_j + u_m + u_n)$$

$$a_2 = \frac{1}{4b}(-u_i + u_j + u_m - u_n)$$

$$a_3 = \frac{1}{4c}(-u_i - u_j + u_m + u_n)$$

$$a_4 = \frac{1}{4bc}(u_i - u_j + u_m - u_n)$$

we have, by proper algebraic substitution and manipulations,

$$u(x, y) = \frac{1}{4bc}[(b - x)(c - y)u_i + (b + x)(c - y)u_j$$
$$+ (b + x)(c + y)u_m + (b - x)(c + y)u_n] \quad (7\text{-}10)$$

Using a similar logical approach, the general displacement function $v(x, y)$ is given by Equation (7-11):

$$v(x, y) = \frac{1}{4bc}[(b - x)(c - y)v_i + (b + x)(c - y)v_j$$
$$+ (b + x)(c + y)v_m + (b - x)(c + y)v_n] \quad (7\text{-}11)$$

c. The element strains are expressed as functions of the general dis-

1. RECTANGULAR PLANE STRESS ELEMENT

placement function $\{\phi\}$. The mathematical relations between strains and displacements are illustrated by the matrix equation (3-5):

$$\{\epsilon\} = \begin{Bmatrix} \epsilon_x \\ \epsilon_y \\ \gamma_{xy} \end{Bmatrix} = \begin{Bmatrix} \dfrac{\partial u}{\partial x} \\ \dfrac{\partial v}{\partial y} \\ \dfrac{\partial u}{\partial y} + \dfrac{\partial v}{\partial x} \end{Bmatrix}$$

When the proper differentiations of the displacement component functions (7-10) and (7-11) are performed, the general displacements, $\{\epsilon\}$ are written in the matrix form as shown in the next page. Equation (7-12) is compatible with expression (4-36).

d. The element stresses are related to the element strains by the general elasticity equation given by (4-39):

$$\{\sigma\} = \begin{Bmatrix} \sigma_x \\ \sigma_y \\ \tau_{xy} \end{Bmatrix} = [D] \begin{Bmatrix} \epsilon_x \\ \epsilon_y \\ \gamma_{xy} \end{Bmatrix}$$

where the elasticity transformation matrix $[D]$ has the form (4-44) for plane stress analysis,

$$[D] = \frac{E}{1 - \nu^2} \begin{bmatrix} 1 & \nu & 0 \\ \nu & 1 & 0 \\ 0 & 0 & \dfrac{1-\nu}{2} \end{bmatrix} \qquad (7\text{-}13)$$

and for plane strain analysis, the elasticity matrix has a form described in expression (4-55):

$$[D] = \frac{E(1-\nu)}{(1+\nu)(1-2\nu)} \begin{bmatrix} 1 & \dfrac{\nu}{1-\nu} & 0 \\ \dfrac{\nu}{1-\nu} & 1 & 0 \\ 0 & 0 & \dfrac{1-2\nu}{2(1-\nu)} \end{bmatrix} \qquad (7\text{-}14)$$

To generate relationships between nodal forces and nodal displacements, the total energy expression is differentiated with respect to the

$$\{\epsilon\} = \begin{Bmatrix} \epsilon_x \\ \epsilon_y \\ \gamma_{xy} \end{Bmatrix}$$

$$= \frac{1}{4bc} \begin{bmatrix} -(c-y) & 0 & (c-y) & 0 & (c+y) & 0 & -(c+y) & 0 \\ 0 & -(b-x) & 0 & -(b+x) & 0 & (b+x) & 0 & (b-x) \\ -(b-x) & -(c-y) & -(b+x) & (c-y) & (b+x) & (c+y) & (b-x) & -(c+y) \end{bmatrix} \begin{Bmatrix} u_i \\ v_i \\ u_j \\ v_j \\ u_m \\ v_m \\ u_n \\ v_n \end{Bmatrix} \quad (7\text{-}12)$$

1. RECTANGULAR PLANE STRESS ELEMENT

displacement components and the results of these differentiation operations are set to be equal to zero. The total potential energy of the system is given by expression (3-1) and also in more detail in expression (4-70). According to the basic minimum-energy principle defined by Equation (3-3), the general equilibrium equations at the nodes are obtained as shown by Equation (4-71):

$$[K]\{\delta\} = \{Q\} \tag{7-15}$$

where $\{\delta\}$ are the unknown nodal displacements, $\{Q\}$ the equivalent nodal loads, and $[K]$ the structural stiffness matrix of the element. Equation (7-15) is rearranged for the direct solution of the nodal displacements, $\{\delta\}$:

$$\{\delta\} = [K]^{-1}\{Q\} \tag{7-16}$$

Equation (7-16) is mathematically analogous to expression (4-72).

Hence the nodal displacements can be obtained as a function of the applied equivalent nodal loads if the stiffness matrix is defined and determined.

The total structural stiffness matrix is obtained by proper superposition of element stiffness matrices as described by Equations (4-69) and (4-80). The general element stiffness matrix $[K]$ is defined by Equation (4-68) as

$$[K] = \int_V \{B\}^T[D]\{B\}\,dV$$

which can be rewritten as

$$[K] = \int \{B\}^T[D]\{B\}t \cdot dx \cdot dy \tag{7-17}$$

where t is the thickness of the plate element. The numerical computations of the element stiffness matrix are performed according to the expressions (4-75) and will have the general form

$$[K] = \begin{bmatrix} K_{ii} & K_{ij} & K_{im} & K_{in} \\ K_{ji} & K_{jj} & K_{jm} & K_{jn} \\ K_{mi} & K_{mj} & K_{mm} & K_{mn} \\ K_{ni} & K_{nj} & K_{nm} & K_{nn} \end{bmatrix} \tag{7-18}$$

Every element matrix of $[K]$ is a 2×2 matrix which makes the dimension of the element matrix 8×8. The integration of expression (7-17) is performed along the limits defined by the dimensions of the rectangular element. The superposition of the element matrices is done according to the pattern defined by Equation (4-80). Once the unknown displacements are computed, they are used in Equation (3-8) to obtain the element strains, and these strains in turn are used to find the element strains by the aid of expression (4-39).

It should be noted that since the assumed general displacement functions are quadratic in nature, the strains in the elements will not be constant.

An engineering evaluation of the rectangular plane stress-plane strain element will be presented at the end of the chapter.

2. TRIANGULAR ELEMENT WITH SIX NODES

To describe more accurately the displacement behavior of a triangular element, the introduction of a quadratic displacement function might be desirable [2, 3]. In practical problems there is a need for such elements, which behave along the mathematical patterns of parabolic functions. The assumption of a full quadratic displacement function requires the use of six coefficients that need to be determined. Since the coefficients are determined by the use of the nodal coordinates, there must be six nodal points associated with the triangular element in order to satisfy the solution conditions. Hence the triangular element with six nodal points becomes one of the genuine finite elements applicable to the analysis. Such an element is shown in Fig. 7-3. The additional three nodes are selected

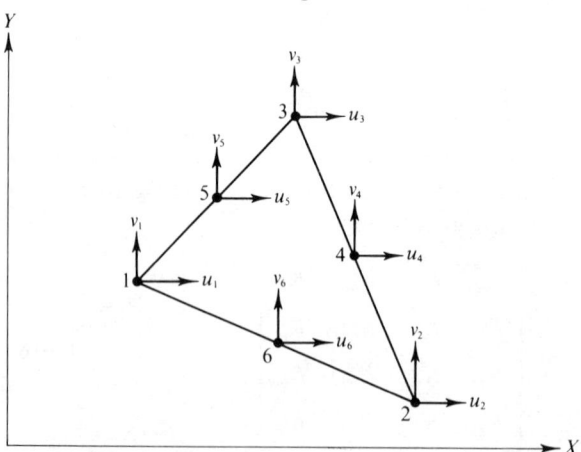

Figure 7-3

to be located at the midpoints of the sides. This configuration will allow the computer programs to compute automatically their coordinates once the corner nodal coordinates are given as input.

2. TRIANGULAR ELEMENT WITH SIX NODES

The displacement compatibility among the adjoining elements must be satisfied. The quadratic function along one side of the triangular element which has three nodes will have a parabolic form. The parabola is well defined by three points on its path. Since the adjacent elements are connected at common joints, their displacement compatibility across the boundaries will be secured. Hence a quadratic form of displacement function is acceptable for finite element analysis. It is important to note that the element strains and stresses will have linear variations in the element which makes it a desirable and flexible type of six noded triangular element.

The logic to be used in the theoretical discussion of the triangular element with six nodes in relation to the finite element method will follow the pattern presented in Chapter 4.

a. The unknowns of the analysis are the nodal displacements $\{\delta\}$ of the element. There are six nodes, and each has two degrees of freedom as the plane stress elasticity conditions prevail.

$$\{\delta\} = \begin{Bmatrix} \delta_1 \\ \delta_2 \\ \delta_3 \\ \delta_4 \\ \delta_5 \\ \delta_6 \end{Bmatrix} = \begin{Bmatrix} u_1 \\ v_1 \\ u_2 \\ v_2 \\ u_3 \\ v_3 \\ u_4 \\ v_4 \\ u_5 \\ v_5 \\ u_6 \\ v_6 \end{Bmatrix} \qquad (7\text{-}19)$$

b. The general displacement functions used have quadratic forms as discussed in the previous paragraphs. These functions are given by

$$\{\phi\} = \begin{Bmatrix} u(x, y) \\ v(x, y) \end{Bmatrix} = \begin{Bmatrix} a_1 + a_2 x + a_3 y + a_4 x^2 + a_5 xy + a_6 y^2 \\ a_7 + a_8 x + a_9 y + a_{10} x^2 + a_{11} xy + a_{12} y^2 \end{Bmatrix} \qquad (7\text{-}20)$$

which, in expanded matrix form, will be written as

$$\{\phi\} = \begin{Bmatrix} 1 & x & y & x^2 & xy & y^2 & 0 & 0 & 0 & 0 & 0 & 0 \\ 0 & 0 & 0 & 0 & 0 & 0 & 1 & x & y & x^2 & xy & y^2 \end{Bmatrix} \begin{Bmatrix} a_1 \\ a_2 \\ a_3 \\ a_4 \\ a_5 \\ a_6 \\ a_7 \\ a_8 \\ a_9 \\ a_{10} \\ a_{11} \\ a_{12} \end{Bmatrix} \quad (7\text{-}21)$$

The determination of the coefficients $a_1, a_2, \ldots, a_{12}$ will require the generation of a system of twelve equations, which are Equations (7-21) in which the coordinates of the six nodal points are substituted. The coordinates of the element are shown in Fig. 7-4. The system of equations developed is given by matrix expression (7-22).

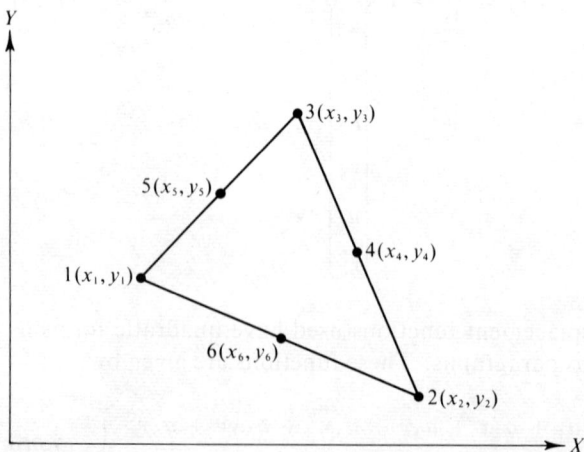

Figure 7-4

2. TRIANGULAR ELEMENT WITH SIX NODES

$$\begin{Bmatrix} u_1 \\ u_2 \\ u_3 \\ \cdot \\ \cdot \\ \cdot \\ u_6 \\ v_1 \\ v_2 \\ \cdot \\ \cdot \\ \cdot \\ v_5 \\ v_6 \end{Bmatrix} = \begin{bmatrix} 1 & x_1 & y_1 & x_1^2 & x_1y_1 & y_1^2 & 0 & 0 & 0 & 0 & 0 & 0 \\ 1 & x_2 & y_2 & x_2^2 & x_2y_2 & y_2^2 & 0 & 0 & 0 & 0 & 0 & 0 \\ & & & & & & & & & & & \\ & & & & & & & & & & & \\ & & & & & & & & & & & \\ & & & & & & & & & & & \\ & & & & & & & & & & & \\ & & & & & & & & & & & \\ & & & & & & & & & & & \\ & & & & & & & & & & & \\ 0 & 0 & 0 & 0 & 0 & 0 & 1 & x_5 & y_5 & x_5^2 & x_5y_5 & y_5^2 \\ 0 & 0 & 0 & 0 & 0 & 0 & 1 & x_6 & y_6 & x_6^2 & x_6y_6 & y_6^2 \end{bmatrix} \begin{Bmatrix} a_1 \\ a_2 \\ a_3 \\ \cdot \\ \cdot \\ \cdot \\ a_6 \\ a_7 \\ a_8 \\ \cdot \\ \cdot \\ \cdot \\ a_{11} \\ a_{12} \end{Bmatrix}$$

(7-22)

Expression (7-22) is rearranged for the direct solution of the coefficients:

$$\begin{Bmatrix} a_1 \\ a_2 \\ a_3 \\ \cdot \\ \cdot \\ \cdot \\ a_{12} \end{Bmatrix} = \begin{bmatrix} 1 & x_1 & y_1 & x_1^2 & x_1y_1 & y_1^2 & 0 & 0 & 0 & 0 & 0 & 0 \\ & & & & & & & & & & & \\ & & & & & & & & & & & \\ & & & & & & & & & & & \\ & & & & & & & & & & & \\ 0 & 0 & 0 & 0 & 0 & 0 & 1 & x_6 & y_6 & x_6^2 & x_6y_6 & y_6^2 \end{bmatrix}^{-1} \begin{Bmatrix} u_1 \\ u_2 \\ u_3 \\ \cdot \\ \cdot \\ \cdot \\ u_{12} \end{Bmatrix}$$

(7-23)

The 12 × 12 matrix is inverted by the aid of a digital computer which allows the expression of the coefficients as the function of nodal displacements. Then the new values of the coefficients are substituted in the general displacement function $\{\phi\}$ given by Equation (7-20). This is similar to expression (4-24) cited in Chapter 4. Therefore, once the nodal displacements are determined, the general displacement functions will also be determined.

c. The element strains are related to the general displacement functions by expression (3-8):

$$\{\epsilon\} = \begin{bmatrix} \epsilon_x \\ \epsilon_y \\ \gamma_{xy} \end{bmatrix} = \begin{bmatrix} \dfrac{\partial u}{\partial x} \\ \dfrac{\partial v}{\partial y} \\ \dfrac{\partial u}{\partial y} + \dfrac{\partial v}{\partial x} \end{bmatrix} \quad (7\text{-}24)$$

The proper differentiation of $\{\phi\}$ will yield the strains as functions of displacements.

$$\begin{Bmatrix} \epsilon_x \\ \epsilon_y \\ \gamma_{xy} \end{Bmatrix} = \begin{bmatrix} 0 & 1 & 0 & 2x & y & 0 & 0 & 0 & 0 & 0 & 0 & 0 \\ 0 & 0 & 0 & 0 & 0 & 0 & 0 & 0 & 1 & 0 & x & 2y \\ 0 & 0 & 1 & 0 & x & 2y & 0 & 1 & 0 & 2x & y & 0 \end{bmatrix} \begin{Bmatrix} a_1 \\ a_2 \\ a_3 \\ \vdots \\ a_{12} \end{Bmatrix} \quad (7\text{-}25)$$

The matrix quantity (7-23) is substituted in expression (7-25) to express the strains as functions of nodal displacements. This will develop a form similar to Equation (4-36)

d. The stresses in the element are related to the strains by an elasticity matrix $[D]$. This relation is given by the matrix equation (4-39), which is the same as Equation (7-26). Depending on the type of stress analysis, either the elasticity matrix quantity (4-44) or (4-55) is used.

$$\{\sigma\} = \begin{Bmatrix} \sigma_x \\ \sigma_y \\ \tau_{xy} \end{Bmatrix} = [D] \begin{Bmatrix} \epsilon_x \\ \epsilon_y \\ \gamma_{xy} \end{Bmatrix} \quad (7\text{-}26)$$

If expressions (7-23) and (7-25) are substituted in Equation (7-26), then the element stresses are expressed as a function of nodal displacements $\{\delta\}$.

e. The nodal displacements are calculated using the general equation (4-71), which is generated by the proper differentiation of the total potential energy of the system with respect to the displacement components and by letting the results of the differentiations be equal to zero. This

2. TRIANGULAR ELEMENT WITH SIX NODES

energy concept is summarized by expression (3-3). The system of equation obtained by this approach are the nodal equilibrium equations.

Equation (4-71) is rearranged for the direct solution of the nodal displacements:

$$\{\delta\} = [K]^{-1}\{Q\} \qquad (7\text{-}27)$$

Equation (7-26) is analogous to Equation (3-11). The matrix $[K]$ is the stiffness matrix of the system and it is obtained by the proper superposition of the element stiffness matrices as defined by expressions (4-69) and (4-80).

The element stiffness matrices are computed from the general expression (4-68) which needs to be integrated, since the $\{B\}$ matrices will include terms with x and y variables:

$$[K]_{\text{element}} = \int_V \{B\}^T [D] \{B\} \, dV \qquad (7\text{-}28)$$

Since the $\{B\}$ matrix is a function of the x and y coordinates, the quantities cannot be removed from the integral sign as was the case for the constant strain triangle. The element stiffness matrix $[K]$ is the 12×12 matrix

$$[K]_{\text{element}} = \begin{bmatrix} K_{11} & K_{12} & K_{13} & K_{14} & K_{15} & K_{16} \\ K_{21} & K_{22} & K_{23} & K_{24} & K_{25} & K_{26} \\ K_{31} & K_{32} & K_{33} & K_{34} & K_{35} & K_{36} \\ K_{41} & K_{42} & K_{43} & K_{44} & K_{45} & K_{46} \\ K_{51} & K_{52} & K_{53} & K_{54} & K_{55} & K_{56} \\ K_{61} & K_{62} & K_{63} & K_{64} & K_{65} & K_{66} \end{bmatrix} \qquad (7\text{-}29)$$

where, each (K_{ij}) is a 2×2 matrix. The element stiffness matrix (7-29) can be expanded using the integrals for (K_{ij}) terms.

$$[K]_{\text{element}} = \begin{bmatrix} \int_V \{B_1\}^T[D]\{B_1\} & \int_V \{B_1\}^T[D]\{B_2\} & \cdots & \int_V \{B_1\}^T[D]\{B_6\} \\ \int_V \{B_2\}^T[D]\{B_1\} & \int_V \{B_2\}^T[D]\{B_2\} & \cdots & \int_V \{B_2\}^T[D]\{B_6\} \\ \vdots & \vdots & \vdots & \vdots \\ \int_V \{B_6\}^T[D]\{B_1\} & \int_V \{B_6\}^T[D]\{B_2\} & \cdots & \int_V \{B_6\}^T[D]\{B_6\} \end{bmatrix}$$

$$(7\text{-}30)$$

where each term is a 2 × 2 matrix as the result of the integrations. The [B] matrix is defined by expressions (4-34) and (4-35).

Then the element matrices defined by Equation (7-30) are superimposed according to expression (4-80) to generate the total structural stiffness matrix $[K]$. Once this matrix quantity is known, the nodal displacements are computed by Equation (7-27), where $\{Q\}$ is the nodal load matrix and $\{\delta\}$ is the nodal displacement matrix.

The strains are computed by Equation (4-36) and finally the element stresses are computed by expression (4-39).

The engineering evaluation of the triangular element with six nodes will be presented at the end of the chapter.

3. PLATE BENDING ELEMENT

The behavior of plates can be analyzed by the finite element method. There are several plate elements which are used in the plate theory [4].

The finite element method discussed here is applied to the analysis of thin plates with small deflections. In thin plates, all stress components are expressed as a function of the vertical displacement expressions of the middle plane. These displacements are related to the X and Y coordinates of the plane of the thin plate. The basic assumptions of the thin-plate theory with small deflections need to be mentioned for the clarity of the presentation [5].

- **a** The geometry of the middle plane defines the actual thin plate and the middle plane keeps its original form during bending.
- **b** Points on a normal to the middle plane before bending occurs stay on the normal to the middle plane after the bending takes place.
- **c** The deflections of the plate are very small compared to its thickness. This classifies it as a thin plate.

The development of a plate bending element presents more mathematical difficulties than those encountered in the discussion of plane stress problems. The satisfaction of the compatibility conditions along the boundaries of adjacent elements as well as within the elements requires the selection of special displacement functions. In some cases additional corrective functions need to be introduced to satisfy the prevailing boundary conditions among the elements [6].

The plate bending elements are usually rectangular or triangular in shape. Each one will require a different kind of general displacement function. Therefore it might be meaningful if each one is discussed separately.

A. Rectangular-Plate Bending Element

A rectangular-plate bending element will have four nodal points and each node will exhibit three degrees of freedom. Figure 7-5 shows such

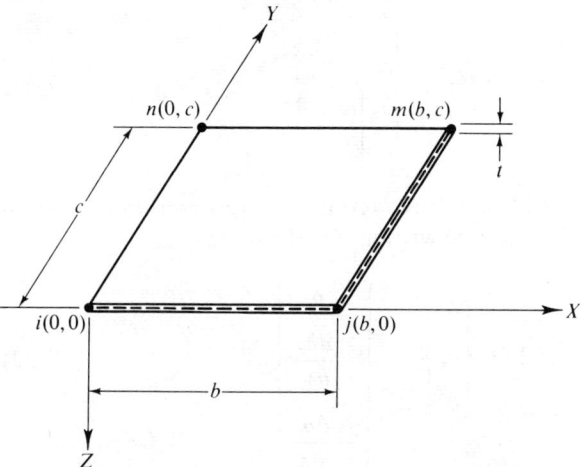

Figure 7-5

an element. This is not a conforming element in that compatibility of normal slopes along the element interfaces may be different. Therefore, convergence with mesh refinement is not ensured.

To illustrate the nodal degrees of freedom the node i is isolated, Fig. 7-6. The right-hand rule is used to define the positive directions of the degrees of freedom in rotation. The three degrees of freedom are: a

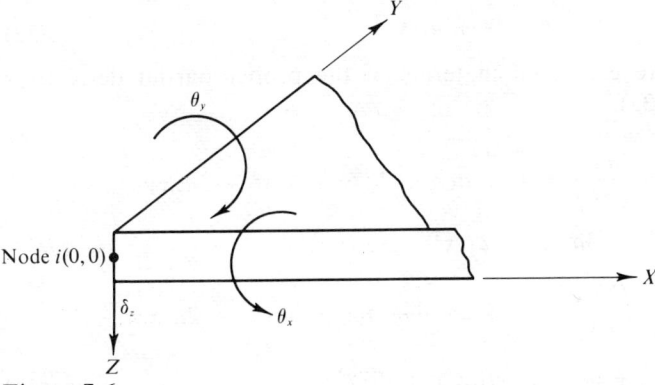

Figure 7-6

displacement in the Z direction, and rotations with respect to the X and Y axes. The rotations can be expressed as the slopes of the vertical displacement, $\{\delta_z\}$.

a. The unknowns of the displacement finite-element method are the nodal displacements. The displacement matrix $\{\delta\}$ for the rectangular plate-bending element has twelve terms:

$$\{\delta\} = \begin{Bmatrix} \delta_i \\ \delta_j \\ \delta_m \\ \delta_n \end{Bmatrix} \tag{7-31}$$

Each term of $\{\delta\}$ will have in turn three terms each representing a degree of freedom for a node. This can be written, for the node i, as

$$\{\delta_i\} = \begin{Bmatrix} \delta_z \\ \theta_x \\ \theta_y \end{Bmatrix} = \begin{Bmatrix} \delta_z \\ \dfrac{\partial \delta_z}{\partial y} \\ -\dfrac{\partial \delta_z}{\partial x} \end{Bmatrix} \tag{7-32}$$

The (θ_y) term is negative due to the reverse direction of the actual rotation with respect to the positive right hand rule.

b. The general displacement function which is assumed to approximate the behavior of the element is a polynomial with twelve terms in x and y [7, 8].

$$\delta_z = a_1 + a_2 x + a_3 y + a_4 x^2 + a_5 xy + a_6 y^2 + a_7 x^3 + a_8 x^2 y$$
$$+ a_9 xy^2 + a_{10} y^3 + a_{11} x^3 y + a_{12} xy^3 \tag{7-33}$$

The rotations are expressed in terms of the proper partial derivatives of expression (7-33).

$$(\theta_x) = \left(\frac{\partial \delta_z}{\partial y}\right) = a_3 + a_5 x + 2a_6 y + a_8 x^2 + 2a_9 xy$$
$$+ 3a_{10} y^2 + a_{11} x^3 + 3a_{12} xy^2 \tag{7-34}$$

$$(\theta_y) = \left(\frac{\partial \delta_z}{\partial x}\right) = a_2 + 2a_4 x + a_5 y + 3a_7 x^2 + 2a_8 xy$$
$$+ a_9 y^2 + 3a_{11} x^2 y + a_{12} y^3 \tag{7-35}$$

3. PLATE BENDING ELEMENT

Expressions (7-33), (7-34), and (7-35) are combined in a matrix form:

$$\begin{Bmatrix} \delta_z \\ \theta_x \\ \theta_y \end{Bmatrix} = \begin{Bmatrix} 1 & x & y & x^2 & xy & y^2 & x^3 & x^2y & xy^2 & y^3 & x^3y & xy^3 \\ 0 & 0 & 1 & 0 & x & 2y & 0 & x^2 & 2xy & 2y^2 & x^3 & 2xy \\ 0 & 1 & 0 & 2x & y & 0 & 3x^2 & 2xy & y^2 & 0 & 3x^2y & y^2 \end{Bmatrix} \begin{Bmatrix} a_1 \\ a_2 \\ a_3 \\ a_4 \\ a_5 \\ a_6 \\ a_7 \\ a_8 \\ a_9 \\ a_{10} \\ a_{11} \\ a_{12} \end{Bmatrix} \quad (7\text{-}36)$$

To determine the coefficients $a_1, a_2, \ldots, a_{12}$, the matrix equation (7-36) is evaluated at each nodal point using the known coordinates values. This will introduce a system of twelve equations. The coordinates are given in Fig. 7-5.

$$\begin{Bmatrix} \delta_{zi} \\ \theta_{xi} \\ \theta_{yi} \\ \delta_{zj} \\ \theta_{xj} \\ \theta_{yj} \\ \delta_{zm} \\ \theta_{xm} \\ \theta_{ym} \\ \delta_{zn} \\ \theta_{xn} \\ \theta_{yn} \end{Bmatrix} = \begin{Bmatrix} 1 & 0 & 0 & 0 & 0 & 0 & 0 & 0 & 0 & 0 & 0 & 0 \\ 0 & 0 & 1 & 0 & 0 & 0 & 0 & 0 & 0 & 0 & 0 & 0 \\ 0 & 1 & 0 & 0 & 0 & 0 & 0 & 0 & 0 & 0 & 0 & 0 \\ 1 & b & 0 & b^2 & 0 & 0 & b^3 & 0 & 0 & 0 & 0 & 0 \\ 0 & 0 & 1 & 0 & b & 0 & 0 & b^2 & 0 & 0 & b^3 & 0 \\ 0 & 1 & 0 & 2b & 0 & 0 & 3b^2 & 0 & 0 & 0 & 0 & 0 \\ 1 & b & c & b^2 & bc & c^2 & b^3 & b^2c & bc^2 & c^3 & b^3c & bc^3 \\ 0 & 0 & 1 & 0 & b & 2c & 0 & b^2 & 2bc & 3c^2 & b^3 & 2bc \\ 0 & 1 & 0 & 2b & c & 0 & 3b^2 & 2bc & b^2 & 0 & 3b^2c & c^2 \\ 1 & 0 & c & 0 & 0 & c^2 & 0 & 0 & 0 & c^3 & 0 & 0 \\ 0 & 0 & 1 & 0 & 0 & 2c & 0 & 0 & 0 & 3c^2 & 0 & 0 \\ 0 & 1 & 0 & 0 & c & 0 & 0 & c^2 & 0 & 0 & c^2 \end{Bmatrix} \begin{Bmatrix} a_1 \\ a_2 \\ a_3 \\ a_4 \\ a_5 \\ a_6 \\ a_7 \\ a_8 \\ a_9 \\ a_{10} \\ a_{11} \\ a_{12} \end{Bmatrix} \quad (7\text{-}37)$$

The expression is rearranged for the direct solution of the coefficients. Once the 12 × 12 matrix is inverted with the aid of a digital computer, the coefficients are expressed as a function of nodal displacements. This states that the determination of the nodal displacements will fix the form of the general displacement polynomial.

c. The strains in a plate element are related to the vertical displacements by the following plate theory expression.[5]

$$\{\epsilon\} = \begin{Bmatrix} -\dfrac{\partial^2 \delta_z}{\partial x^2} \\ -\dfrac{\partial^2 \delta_z}{\partial y^2} \\ 2\dfrac{\partial^2 \delta_z}{\partial x \partial y} \end{Bmatrix} \quad (7\text{-}38)$$

Substituting of expressions (7-34) and (7-35) in Equation (7-38) and then performing the differentiation will yield more detailed matrix equations for the element strains.

$$\frac{\partial^2 \delta_z}{\partial x^2} = 2a_4 + 6a_7 x + 2a_8 y + 6a_{11} xy \quad (7\text{-}39)$$

$$\frac{\partial^2 \delta_z}{\partial y^2} = 2a_6 + 2a_9 x + 6a_{10} y + 6a_{12} xy \quad (7\text{-}40)$$

$$\frac{\partial^2 \delta_z}{\partial x \partial y} = a_5 + 2a_8 x + 2a_9 y + 3a_{11} x^2 + 3a_{12} y^2 \quad (7\text{-}41)$$

Then expression (7-38) is rewritten as (7-42), using the functions (7-39), (7-40), and (7-41).

$$\{\epsilon\} = \begin{pmatrix} 0 & 0 & 0 & -2 & 0 & 0 & -6x & -2y & 0 & 0 & -6xy & 0 \\ 0 & 0 & 0 & 0 & 0 & -2 & 0 & 0 & -2x & -6y & 0 & -6xy \\ 0 & 0 & 0 & 0 & 2 & 0 & 0 & 4x & 4y & 0 & 6x^2 & 6y^2 \end{pmatrix} \begin{Bmatrix} a_1 \\ a_2 \\ a_3 \\ a_4 \\ a_5 \\ a_6 \\ a_7 \\ a_8 \\ a_9 \\ a_{10} \\ a_{11} \\ a_{12} \end{Bmatrix}$$

(7-42)

If the values of the coefficients which are expressed in Equation (7-37) are substituted in Equation (7-42), then the strains will be direct functions of nodal displacements.

d. The stresses are related to the strains by an elasticity matrix $[D]$, which has different forms for various types of elasticity problems. The elasticity relationship is given by expression (4-39).

$$\{\sigma\} = \begin{Bmatrix} \sigma_x \\ \sigma_y \\ \tau_{xy} \end{Bmatrix} = [D] \begin{Bmatrix} \epsilon_x \\ \epsilon_y \\ \gamma_{xy} \end{Bmatrix} \qquad (7\text{-}43)$$

For the case of an elastic *isotropic* plate, the elasticity matrix is defined by Equation (7-44) [5].

$$[D] = \frac{Et^3}{12(1-\nu^2)} \begin{Bmatrix} 1 & \nu & 0 \\ \nu & 1 & 0 \\ 0 & 0 & \frac{1-\nu}{2} \end{Bmatrix} \qquad (7\text{-}44)$$

Where E is Young's modulus, t is the plate thickness, and ν is the Poisson ratio.

It needs to be noted that the strain and stress values vary linearly across the plate thickness. The values of the stresses at a given point are determined, using the known beam relation.

$$\sigma_x = \frac{M_x \cdot z}{I} \qquad (7\text{-}45)$$

where σ_x is a stress component z units from the middle plane, M_x the bending moment about the x axis, I is moment of inertia of the section, and z the distance where the stress σ_x is computed.

For a rectangular cross section of width unity and height t, the moment of inertia is $(t^3/12)$. The other internal moments are M_y and M_{xy}, which are defined by Fig. 7-7.

e. The unknown displacements $\{\delta\}$ are computed by the solution of the general equation defined by Equation (4-71) which is repeated for the continuity of the discussion:

$$[K]\{\delta\} = \{Q\} \qquad (7\text{-}46)$$

Expression (7-46) is obtained by the application of principle of stationary total potential energy. The partial derivatives of the total energy with respect to the nodal displacements, when set to be equal to zero, generate the nodal equilibrium equations. The coefficients of this system of equations relate the nodal loads to the nodal displacements and they form,

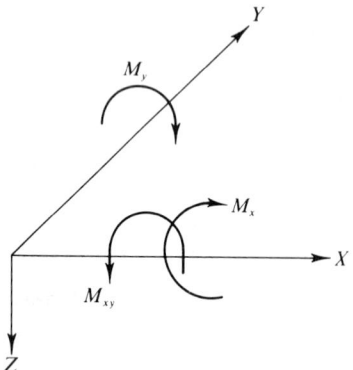

Figure 7-7

when written in a matrix form, the structural stiffness matrix. At each node the three possible loads are two moments and one vertical load.

The development of the stiffness matrix is according to the general expression (4-69).

$$[K] = \int_V \{B\}^T[D]\{B\} \, dV \tag{7-47}$$

If the plate thickness t is constant, expression (7-47) can be written as a surface integral:

$$[K] = t \cdot \int \{B\}^T[D]\{B\} \, dxdy \tag{7-48}$$

The expression under the integral sign can be expanded and then the integration operations can be performed. This yields the stiffness matrix of one element. After the stiffness matrices of all elements are determined, they are properly superimposed according to the pattern defined by the matrix form, Equation (4-80).

The total stiffness matrix so determined is used in Equation (7-46) to obtain the unknown nodal displacements.

$$\{\delta\} = [K]^{-1}\{Q\} \tag{7-49}$$

To complete the discussion, the results of the equation are substituted in Equation (4-36) to compute the element strains. Then the stresses are computed by substituting the known strains in Equation (4-39).

COMMENTS ON THE NATURE OF DISPLACEMENT FUNCTION

The general displacement polynomial defined by expression (7-33) has twelve terms. It reduces to four terms along the boundaries of the rectangular plate elements due to the constant value of one of the variables. Mathematically it has a cubic form with four coefficients to be determined. The two nodes of the boundary exhibit four conditions (two displacements and two slopes) which are sufficient to define the boundary. In plate-bending analysis the conditions to be considered are the continuity of displacements, slopes, and second derivatives between adjacent elements.

The same thing cannot be said about the compatibility of the slopes between elements. The general slopes expressions along the two boundaries are given by expressions (7-34) and (7-35). Along the boundaries they become parabolic in nature. The parabola is uniquely defined by three points. Unfortunately, the boundaries can furnish only two usable conditions which are the normal slopes at the two nodal points. Therefore the compatibility of slopes across the boundaries of the elements is not secured. To remedy this, corrective functions might be introduced into the analysis problem [6]. However, the assumed 12-term polynomial displacement function might yield acceptable results to the analysis problems.

B. Triangular-Plate Bending Element

A triangular-plate element will have three nodes and each node will have three degrees of freedom—namely, one vertical displacement and two rotations. An element is shown in Fig. 7-8. To illustrate the degrees of freedoms, the node i is isolated as given by Fig. 7-9.

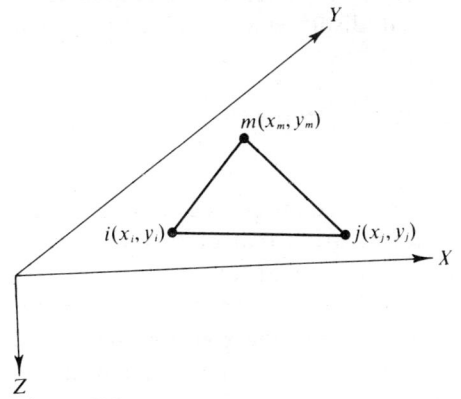

Figure 7-8

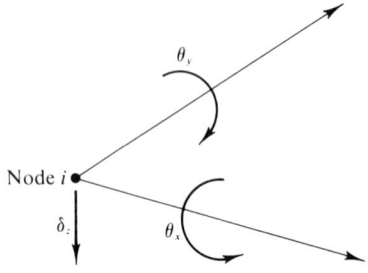

Figure 7-9

a. The unknowns of the analysis are the nodal displacements. There are nine possible displacements, three at each node. For the node i, expression (7-49) is written similar to Equation (7-32).

$$\{\delta\}_i = \begin{Bmatrix} \delta_z \\ \theta_x \\ \theta_y \end{Bmatrix} \begin{Bmatrix} \delta_z \\ -\dfrac{\partial \delta_z}{\partial y} \\ \dfrac{\partial \delta_z}{\partial x} \end{Bmatrix} \quad (7\text{-}50)$$

The total displacement matrix $\{\delta\}$ will have nine matrix elements:

$$\{\delta\} = \begin{Bmatrix} \delta_i \\ \delta_j \\ \delta_m \end{Bmatrix} \quad (7\text{-}51)$$

b. The general displacement function is assumed to be a cubic expansion with ten terms [9]. However, two of the terms are combined in order to match the actual nine nodal displacements of a triangular element.

$$\delta_z = a_1 + a_2 x + a_3 y + a_4 x^2 + a_5 xy + a_6 y^2 + a_7 x^3 \\ + a_8(x^2 y + xy^2) + a_9 y^3 \quad (7\text{-}52)$$

The displacement field defined by the above equation does not produce a conforming element and results in a formulation that does not converge with mesh refinement. An important problem with this function is that the matrix used to determine the coefficients a_i becomes singular when the two sides of the triangular element are parallel to the x and y axes [6]. If expression (7-51) is substituted in Equation (7-49), the nodal displacements can be written as function of the coefficients:

COMMENTS ON THE NATURE OF DISPLACEMENT FUNCTION 179

$$\begin{Bmatrix} \delta_z \\ \theta_x \\ \theta_y \end{Bmatrix} = \begin{bmatrix} 1 & x & y & x^2 & xy & y^2 & x^3 & (x^2y + xy^2) & y^3 \\ 0 & 0 & -1 & 0 & -x & -2y & 0 & -(x^2 + 2xy) & -3y^2 \\ 0 & 1 & 0 & 2x & y & 0 & 3x^2 & (2xy + y^2) & 0 \end{bmatrix} \begin{Bmatrix} a_1 \\ a_2 \\ a_3 \\ a_4 \\ a_5 \\ a_6 \\ a_7 \\ a_8 \\ a_9 \end{Bmatrix}$$

(7-53)

To express the coefficients $a_1, a_2, \ldots, a_9$ as a function of the nodal displacements, the actual nodal coordinates of each node needs to be substituted in Equation (7-52). This process will generate nine equations with nine unknowns as given by the matrix equation (7-54).

$$\begin{Bmatrix} \delta_{zi} \\ \theta_{xi} \\ \theta_{yi} \\ \delta_{zj} \\ \theta_{xj} \\ \theta_{yj} \\ \delta_{zm} \\ \theta_{xm} \\ \theta_{ym} \end{Bmatrix} = \begin{bmatrix} 1 & x_i & y_i & x_i^2 & x_iy_i & y_i^2 & x_i^3 & (x_i^2y_i + x_iy_i^2) & y_i^3 \\ 0 & 0 & -1 & 0 & -x_i & -2y_i & 0 & -(x_i^2 + 2x_iy_i) & -3y_i^2 \\ 0 & 1 & 0 & 2x_i & y_i & 0 & 3x_i^2 & (2x_iy_i + y_i^2) & 0 \\ 1 & x_j & y_j & x_j^2 & x_jy_j & y_j^2 & x_j^3 & (x_j^2y_j + x_jy_j^2) & y_j^3 \\ 0 & 0 & -1 & 0 & -x_j & -2y_j & 0 & -(x_j^2 + 2x_jy_j) & -3y_j^2 \\ 0 & 1 & 0 & 2x_j & y_j & 0 & 3x_j^2 & (2x_jy_j + y_j^2) & 0 \\ 1 & x_m & y_m & x_m^2 & x_my_m & y_m^2 & x_m^3 & (x_m^2y_m + x_my_m^2) & y_m^3 \\ 0 & 0 & -1 & 0 & -x_m & -2y_m & 0 & -(x_m^2 + 2x_my_m) & -3y_m^2 \\ 0 & 1 & 0 & 2x_m & y_m & 0 & 3x_m^2 & (2x_my_m + y_m^2) & 0 \end{bmatrix} \begin{Bmatrix} a_1 \\ a_2 \\ a_3 \\ a_4 \\ a_5 \\ a_6 \\ a_7 \\ a_8 \\ a_9 \end{Bmatrix}$$

(7-54)

The equation can be rearranged in order to obtain the coefficients directly once the displacements are determined.

  c. The strains are related to the vertical displacements by the following elasticity relationship,

$$\{\epsilon\} = \left\{ \begin{array}{c} -\dfrac{\partial \delta_z^2}{\partial x^2} \\[6pt] -\dfrac{\partial \delta_z^2}{\partial y^2} \\[6pt] 2\dfrac{\partial^2 \delta_z}{\partial x \partial y} \end{array} \right\} \qquad (7\text{-}55)$$

Substituting Equation (7-51) in Equation (7-54) and performing the necessary differentiation, one obtains the matrix equation

$$\{\epsilon\} = \begin{Bmatrix} 0 & 0 & 0 & -2 & 0 & 0 & -6x & -2y & 0 \\ 0 & 0 & 0 & 0 & 0 & -2 & 0 & -2x & -6y \\ 0 & 0 & 0 & 0 & 2 & 0 & 0 & 4(x+y) & 0 \end{Bmatrix} \begin{Bmatrix} a_1 \\ a_2 \\ a_3 \\ a_4 \\ a_5 \\ a_6 \\ a_7 \\ a_8 \\ a_9 \end{Bmatrix}$$

(7-56)

d. The element stresses are related to the element strains by the elasticity matrix [D] which is defined, for an isotropic plate, by the matrix expression (7-44). The governing equation is Equation (7-43).

$$\{\sigma\} = [D]\{\epsilon\} \qquad (7\text{-}57)$$

The rest of the analysis related to the triangular plate-bending element will follow the same logic used for the discussion of the rectangular plate bending element. Therefore the present discussion will terminate here. The displacement and slope compatibilities across the boundaries have the pattern presented by the behavior of the rectangular-plate bending element.

COMPARISON OF THE VARIOUS ELEMENTS

A descriptive comparison of the five types of finite elements presented in the book might be of practical value to the reader. It is obvious each one of the elements has particular characteristics which could be assets for the analysis of typical problems. They are all used in practice and are incorporated in many finite element computer programs.

i. Constant-Strain Triangular Element

This is the first element extensively employed in engineering analysis projects. It is used for the analysis of plane stress and plane strain elasticity problems. The general displacement function associated with it is a linear function which secures the continuity between the elements. Almost all the digital computer programs based on the finite element method have this element in their subroutines. "Z-C," ELAS, ICES-STRUDL, NASTRAN, FEPLS, and Wilson are a few of the available programs which have the constant-strain triangular elements in their logic. The triangular torus element used in the analysis of axisymmetric solids obeys similar rules of the constant-strain triangle.

ii. Constant-Strain Rectangular Element

The theoretical relations associated with the rectangular element are similar to the ones of triangular element. The general displacement function is a four-termed polynomial, since the rectangle has four nodes. For structural systems in plane stress or plane strain with regular boundaries, the rectangular constant strain element is useful. However, as the geometry of the structure becomes complex, the rectangle cannot efficiently describe its boundaries. This limits the practical applications of the constant-strain rectangular element.

iii. Triangular Element with Six Nodes

Since the triangular element introduced here has six nodes, then it is mathematically justified to use a full quadratic polynomial for the general displacement function of the element. The form of this function secures linear variations of the stresses and strains in the element while obeying plane stress elasticity principles. This is a refined element with respect to the constant-strain triangular and rectangular elements. It fits complex boundaries with ease. Its applications are a great asset to the engineer.

Computer programs such as STRUDL, NASTRAN, and ELAS have this element in their subroutines.

iv. Rectangular-Plate Bending Element

The rectangular-plate bending element is one of the basic bending elements. It has four nodal points and each node has three degrees of freedom—a lateral displacement and two rotations. The availability of 12-nodal displacement components allows the introduction of a general displacement polynomial with twelve terms. The stresses and strains will be quadratic in mathematical form, approaching the exact behavior of the plate. The vertical displacements along the boundaries are satisfied, since a cubic equation is uniquely defined by four given values—two displacements and two rotations. However, normal slope compatibility is not satisfied, because the variation of the slope along the boundary is parabolic and there exists only two boundary values, the two normal slopes at the nodes. A parabola is uniquely defined by three known points. Hence the slope compatibility is not satisfied. However, this approach still yields acceptable results for the analysis of plate-bending elements. There are possible corrective measures to secure better compatibility of the slopes across the element boundaries [6]. Programs such as NASTRAN, STRUDL, and ELAS have this type of element in their subroutines.

v. Triangular-Plate Bending Element

The triangular-plate bending element has an advantage over the rectangular-plate bending element due to its adaptability to describe the existing boundaries with efficiency and ease. The assumed displacement function is a full cubic polynomial with ten terms. This introduces a mathematical difficulty in the determination of the ten constants, for there are only nine nodal displacement components associated with the triangular bending element. To overcome this difficulty several schemes are used, such as the omission of one term or combination of two similar terms. Even though these will be arbitrary, acceptable solutions are obtained. The tenth constant in the triangular finite element in bending can be determined from the minimum strain energy [10].

The displacement compatibility across the boundaries is satisfied, but the compatibility of the normal slopes at the boundaries is not. The reasons for this are similar to those given in the discussion of the rectangular plate-bending element. Digital computer programs such as NASTRAN, STRUDL, and ELAS use this element in their approach to the analysis of plate-bending problems. There are other types of finite elements such as those mentioned during the presentation of the ELAS computer programs. The three-dimensional elements such as tetrahedron elements are used for the analysis of three-dimensional elasticity problems. More information is available on three-dimensional finite element analysis in various engineering journals [11–13].

REFERENCES

1. R. W. CLOUGH, "The Finite Element in Plane Stress Analysis," Proceedings of 2nd ASCE Conference on Electronic Computations, Pittsburgh, Pa., September 1960.
2. FRAEIJS DE VEUBEKE, B. "Displacement and Equilibrium Models in the Finite Element Method," in O. C. Zienkiewicz and G. S. Holister (eds.), 9, *Stress Analysis*. New York: Wiley, 1965.
3. J. H. ARGYRIS, "Triangular Elements with Linearly Varying Strain for the Matrix Displacement Method," *Journal of Royal Aeronautical Society Technical Note* **69**, 711–713 (October 1965).
4. R. W. CLOUGH and J. L. TOCHER, "Finite Element Stiffness Matrices for Analysis of Plate Bending," Proceedings of the Conference on Matrix Methods of Structural Analysis, Wright-Patterson Air Force Base, Ohio 515–546. (October 26–28, 1965).
5. S. TIMOSHENKO and S. WOINOWSKY-KRIEGER, *Theory of Plates and Shells*. New York: McGraw-Hill, 1959.
6. O. C. ZIENKIEWICZ and Y. K. CHEUNG, *The Finite Element Method in Structural and Continuum Mechanics*. New York and London: McGraw-Hill, 1967, p. 116.
7. A. ADINI and R. W. CLOUGH, "Analysis of Plate Bending by the Finite Element Method," Report submitted to the National Science Foundation, 1960.
8. R. J. MELOSH, "Basis for Derivation of Matrices for the Direct Stiffness Method," *AIAA Journal* **1** (1963).
9. J. L. TOCHER, "Analysis of Plate Bending Using Triangular Elements," Ph.D. Thesis, University of California, Berkeley, 1962.
10. S. UTKU, "Computation of Stresses in Triangular Finite Elements," Technical Report No. 37-948, Jet Propulsion Laboratory, Pasadena, California, 1966.
11. R. H. GALLAGHER, J. PADLOG, and P. P. BIJLAARD, "Stress Analysis of Heated Complex Shapes," *Journal of Aero-Space Science*, 1962.
12. R. J. MELOSH, "Structural Analysis of Solids," *Proceedings of the American Society of Civil Engineers*, S.T. **4**, 205–223 (August 1963).
13. J. H. ARGYRIS, "Matrix Analysis of Three Dimensional Elastic Media—Small and Large Displacements," *AIAA Journal* **3** (January 1965).

Computer Programs Based on the Finite Element Method

INTRODUCTION

The practical value and success of the finite element method as we have noted, is associated with the availability and use of high-speed electronic digital computers. Another asset of the finite element method stems from its adaptability to describe the complex geometry and boundary conditions of analysis problems and its ability to produce rather quick and accurate results. The formulation of all types of analysis problems is done by matrix quantities, and matrix operations have to be performed to reach accurate results. The necessity of incorporating the use of digital computers in finite element analysis caused the development of several computer programs with various capabilities. There are large, general-purpose computer programming systems which are able to solve a wide variety of problems; then there are comparatively small specialized computer programming systems which are able to analyze particular types of problems.

GENERAL-PURPOSE AND SPECIAL-PURPOSE PROGRAMS

The general-purpose programs offer many practical qualities.

- **a** They are *user-oriented*, meaning that the input quantities are comparatively easy to prepare and they are not to be written in special sophisticated computer languages. An engineer does not need to be familiar with computer hardware or software to learn how to use the general-purpose programming system.
- **b** They are very large computer programming systems which are able to solve many types of simple or complex problems.
- **c** They are modular in nature which allows the addition of new sections to the main program. This is a desirable quality, since it permits the continuous improvement and extension of the program. General-purpose programs may be called *dynamic systems*, since they are open to positive change.
- **d** They are readily available at most computer centers associated with educational and research institutions as well as at major consulting firms. Since their developments are usually sponsored by government or wealthy corporations, the user does not pay their initial cost. Hence, they are comparatively cheap to use.

Unfortunately, there are some disadvantages related to the general-purpose programs. They must also be mentioned for their proper evaluation.

- **a** The initial cost of the development of general-purpose programs is very high.
- **b** The size of these programs decreases computer efficiency. Many unnecessary logical steps might be performed due to the nature of subroutines, even though the problem is simple.
- **c** The internal complexity of the software makes it almost impossible for a user to make changes in the logic of the program. The programs must be used as they are initially written and stored in the computer.
- **d** They usually require a computer system with a comparatively large storage capacity.

The general-purpose programs are efficient communication devices for the engineers and free them from the burden of tedious operations.

Special-purpose programs present another approach to the programming and the use of digital computers. They have several qualities that make them attractive.

a   Special purpose programs are short programs compared to general-purpose programs. Hence, their initial cost is not high.
b   They require smaller computers due to their conservative size. Smaller computer centers will prefer operating using these types of programming systems.
c   Since they are simpler and shorter programs, it is possible to alter and expand these programs without excessive work.
d   They have high efficiency in solving problems of moderate size and complexity.

The major disadvantage of special-purpose programs is their lack of capability in solving various types of problems. They are prepared for the analysis and design of specialized problems. A new special-purpose program is needed for nearly every special problem. They are not usually written as user-oriented programs, and present some difficulties in the preparations of their input quantities as well as the understanding of their abilities and operations.

There are general-purpose programs and special-purpose programs which are based on the finite element method of analysis. They will be discussed, applied, and evaluated in Chapters 9 and 10.

COMPUTER PROGRAMS

A few of the computer programs based on the principle of the finite element method need to be mentioned. They are an integral part of the success of this method.

1   ICES-STRUDL subprogram (Integrated Civil Engineering System-Structural Design Language)
2   NASTRAN Program (NASA Structural Analysis program)
3   SAAS-II, (Structural Analysis of Axi-symmetric Solids)
4   ELAS Program (ELAsticity)
5   STRATA (STRess And Thermal Analysis)
6   Zienkiewicz Program
7   Wilson's Programs
8   BOSOR (Buckling Of Shells Of Revolution)
9   SLADE Program
10 FEPLS Program (Finite Element analysis of Plates)

A more complete list of these programs can be found in the publications of the Illinois Institute of Technology Research Institute.

There is no intention of mentioning all the programs here, since it is almost impossible to be aware of all of them. This is a young scientific field that is continuously growing. The author will discuss those he has had the opportunity to become familiar with and has used in his research and teaching activities. Their presentation will be more of a descriptive nature than a complete and detailed discussion. Most of them have their documentation and user manuals, comparable in length to a textbook. When detailed and comprehensive information about these programs are needed, their original documentations should be consulted.

1. ICES-STRUDL SUBPROGRAM [1]

The Integrated Civil Engineering system-STRuctural Design Language is discussed in Chapter 9 in detail with several applications on how to analyze problems. This is a large general-purpose computer programming system which is available at the software libraries of the majority of computer centers. It was developed at Massachusetts Institute of Technology with the financial backing of International Business Machines Corporation.

STRUDL-Version II has finite-element capability to treat one-, two-, and three-dimensional structural analysis problems [2]. It has more than 40,000 Fortran statements. STRUDL can solve plane stress and plane strain elasticity problems, plate-bending problems, shallow-shell structures, and linear plane structural systems. The commonly used elements are

  Constant-strain triangle—global formulation: CSTG
  Constant-strain triangle—local formulation: CSTL
  Linear-strain triangle: LST
  Plane stress and strain rectangle: PSR
  Flat-plate triangle: CPT
  Flat-plate rectangle: BPR
  Flat-plate parallelogram: BPP
  Shallow-shell curved rectangle: SSCR

The efficiency of STRUDL depends on the size of the computer hardware and also on the nature of the analysis problem. Since Chapter 9 deals with STRUDL, the reader is referred to that particular chapter for more relevant information. STRUDL can be ordered from Bechtel Engineering Corporation, San Francisco, California.

2. NASTRAN PROGRAM [3]

NASTRAN stands for NASA STRuctural ANalysis program, developed by the National Aeronautics and Space Administration. It is a large general-purpose digital computer program able to analyze elastic structural systems using the displacement finite-element method. NASTRAN can solve any type of linear structure made of beams, linear elements, shear and twist panels, plates, and solids of revolution. The algorithms used for the development of the program can handle very large problems with many degrees of freedom for nodal displacements. It can solve one-, two-, and three-dimensional problems. Its efficiency increases with the size of the structural system considered. In all computer calculations double precision is used. NASTRAN is adaptable to different computer systems. Table 8-1 illustrates the relationships between the computer hardware and the program [4].

The program costs from $1400 to $1800, depending on the choice of options. It is available from Computer Software Management and

TABLE 8-1

Computer Hardware	Operating System	Minimum Main Core Assignment	Required Peripheral Storage
IBM 7094/7040 direct coupled (available at level 8.0 only)	IBSYS/DCOS	32K words	One 1301 disc module and tapes
UNIVAC 1108	EXEC 8	42K words	At least one FASTRAND drum
IBM 360 model 50	OS 360/PCP	200K bytes of a 256K-byte core	At least one 2314 disc
IBM 360 models 50, 65, 67	OS 360/PCP or MFT HASP permitted	300K bytes of a 512K-byte core	At least one 2314 disc
IBM 360 models 50, 65, 67, 75 85, 91, 95	OS 360/PCP or MFT or MVT	400K bytes of a 1000K-byte core	At least one 2314 disc
CDC 6400 6500 6600	SCOPE 3.2 RUN compiler with nonstandard returns	130K octal	3 million words of 6803 disc

Information Center (COSMIC), University of Georgia, Athens, Ga. Since NASTRAN is a large programming system and comparatively new, it is not readily available in computer centers yet, however, its possibilities for wide application are very great.

3. SAAS-II PROGRAM [5]

SAAS-II stands for finite element stress Analysis of Axisymmetric solids with orthotropic, temperature-dependent material properties. It deals with axisymmetric solids under axisymmetric loading conditions. The loads that can be considered are thermal forces, surface pressures, axial acceleration forces, and nodal forces. Triangular or quadrilateral torus elements are used for the idealization of the continuous structural system. SAAS-II can accept variations in material properties as well as complexities in the geometry. One of the desired characteristics of the program is its ability to generate automatically a mesh for the problem. This is a time-saver operation for the user, and it also decreases the possibility of introducing errors in defining the topology of the structural system. It uses a Gaussian elimination procedure for the solution of the system of linear equations to determine the unknown nodal displacements $\{\delta\}$. The output can be printed and plotted for quick engineering evaluation.

SAAS-II is written in Fortran IV language. The system has a main program and 20 subroutines which define the steps of the development of finite-element method as presented in Chapter 3. CDC series require single precision since they have 10-bit word lengths. IBM 360 series require double precision for computer calculations, since they have only 4-bit word lengths. The convergence concept has been proven to be true for SAAS-II as the element sizes decrease.

It is written to handle up to 1000 nodal points, 1000 elements, 6 different materials, 12 temperatures and 200 different pressures. The program in its original form should be used in a computer with 216K core storage (IBM 360) or in a computer with 142K core storage (CDC 6600).

This is a special purpose program for the analysis of axisymmetric solids. The theory behind the analysis of axisymmetric solids by the finite-element method has been presented in Chapter 6.

The program can be ordered from Clearinghouse, U.S. Department of Commerce, Springfield, Virginia. The two major characteristics of SAAS-II are its ability to generate the mesh required by the analysis procedure and plot the obtained results.

4. ELAS PROGRAM [6]

ELAS stands for the first two syllables of the world *elas*ticity. It is a general-purpose computer program for the equilibrium problems of linear structures. It was originally written in Fortran II computer language. ELAS is theoretically based on the displacement finite element method. It can handle a wide variety of analysis problems in one-, two-, and three-dimensional spaces. It is a flexible system due to its large element library. It can generate a mesh using lineal, triangular, quadrilateral, tetrahedral, hexahedral, conical, triangular torus, and quadrilateral torus elements. ELAS is known for its efficient computer operations because of its specialized logical interrelations between the subroutines. The source deck has about 8000 cards and the object deck includes about 1400 binary cards. ELAS can handle up to 99 variations of the following quantities:

Materials
Temperature changes
Temperature gradients
Moment of inertia
Cross-sectional areas
Thicknesses
Pressures
Torsional constants

The number of finite elements and nodal points in the mesh may reach 9999. The degrees of freedom that can be incorporated in the solution of the problem depend on the size of the computer hardware. A 32K machine can handle 600 degrees of freedom, while a 64K computer can handle 1500 degrees of freedom. The general input quantities include the coordinates of the nodes, the geometric dimensions, topology, material properties, loading characteristics, and prescribed forces and/or displacements at nodal points. The general output quantities consist of the listing of the input information, displacements at nodal points, and stresses at nodal points.

ELAS uses a special Cholesky scheme for the solution of the system of linear equations generated by the equilibrium conditions at the nodes. The Cholesky algorithm is an asset since it decreases the storage space requirements as well as the steps of matrix operations. The scope of ELAS can be well defined by mentioning the types of problems it can analyze: plane and space trusses, plane and space frames, gridwork systems, plane stress and plane strain, plate bending, general solids, membrane and bending analysis of general shells, solids of revolution, and shells of revolution. Even though it is a general-purpose program, its size is conservative allowing its use in a variety of computer systems. ELAS is distributed through the Computer Software Management and Information Center, (COSMIC), University of Georgia, Athens, Ga.

5. STRATA PROGRAM [7]

STRATA is a symbolic abbreviation for STRess And Thermal Analysis. It is intended for elastic-plastic stress analysis of plane stress and axisymmetric problems. It can be classified as a special-purpose program, since it is related to plane stress and axisymmetric structural analysis. It is written in Fortran IV computer language. The source deck consists of about 2000 cards. Originally it was prepared to run on the IBM 360-50 digital computer.

STRATA has thirteen subroutines whose operations complement the program:

FETS01 Reads input and performs data-error analysis.
FETS02 Performs temperature analysis.
FETS03 Performs stress analysis.
FETS04, FETS05, FETS06, FETS07 Perform various matrix operations.
FETS09 Modifies stiffness matrix for displacement boundary conditions.
FETS10 Solves the system of linear equations by matrix reduction.
FETS11 Calculates strains and stresses.
FETS12 Gaussian quadrature over triangular region.
FETS13 Calculates portion of element load matrix due to $\{\epsilon_0\}$.

STRATA uses quadratic expansion for the nodal displacements which causes linear strain variations across the elements. This approach requires fewer numbers of elements to reach an acceptable solution than when constant-strain elements are employed for the analysis. The plastic-stress analysis part of the program uses an incremental loading procedure and the Prandtl-Reuss equations, [8], to reach a solution.

The finite element logic is similar to the one presented in Chapter 3. Once the displacements are determined, strains and stresses are computed. The stress values are given at nodal points as well as at the centroid of the elements.

STRATA can use a subprograms, "Data Check Programs and Plots," to check the following input data to secure a correct output.

a Finite element mesh and element numbers
b Nodal point numbers
c Material boundaries
d Voids and overlaps in the topology of the system

This program can be ordered from the Department of Engineering Mechanics, University of Missouri-Rolla.

6. ZIENKIEWICZ-CHEUNG PROGRAM [9]

The "Z-C" Program is a special-purpose program for the analysis of plane stress and plane strain elasticity problems. It uses constant-strain triangular elements to generate the mesh of the structural system. It is written in Fortran IV computer language and its source deck consists of about 450 cards. The program is controlled by a main program which directs and coordinates four subroutines.

The logic followed for the development of this program is similar to that discussed in Chapter 3:

Read input data.
Print input data.
Calculate the applied loads.
Generate element stiffness matrices.
Assemble the global stiffness matrix.
Introduce the known displacements of nodes.
Compute the unknown nodal displacements.
Compute element strains.
Compute element stresses.
Compute principal stresses.

The node numbering needs to be performed in the narrow dimension of the structure to insure operational efficiency. A large structure can be partitioned and analyzed separately and then assembled. This capability of the "C-K" program will allow the use of a comparatively small computer to solve actual problems.

The output includes nodal displacements, coordinates of the centroids of triangular elements, the stresses at the centroids of the elements, the principal stresses at the centroids of the elements, and the value of the principal angle for each element. This program is included in the text by Zienkiewicz and Cheung [9].

7. WILSON PROGRAM [10]

There are several finite element computer programs developed by Dr. E. L. Wilson. The one mentioned here is the computer program related to his Ph.D. thesis completed in 1963. The program handles linear elastic structures using triangular elements. In its original form it would analyze only plane stress problems. However, with slight variation in elasticity matrix it can be applied to the analysis of plane strain problems.

It goes through three major operations: the development of the system of equilibrium equations, the solution of this system for the unknown nodal displacements, and calculations of element stresses. The size of the analysis problem that can be solved by this program depends on the capabilities of the available digital computer. For example, a structural system with 550 triangular elements and 340 nodal points can be analyzed on a computer with 32K storage capacity. A copy of this special purpose program can be obtained from University Microfilms, Inc., Ann Arbor, Mich.

8. BOSOR-3 PROGRAM [11]

This is a computer program to perform stress, stability and vibration analysis of complex shells of revolution. The basic assumptions of BOSOR-3 are: the material is elastic, the theory of thin shells is governing, and the structural system is axisymmetric.

To generate the mesh for the idealization of the shell structures, shell segments are used as finite elements. The capacity of BOSOR is flexible, since it considers the shell segment elements individually. The program can handle up to 25 shell segments. Each segment can have a maximum of 97 nodal points. The structural system itself can have a maximum of 450 nodes. This limitation is dictated by the core size of the UNIVAC 1108 computer. If the IBM 360 operating system is used, double precision will be necessary to obtain acceptable results. The BOSOR-3 manual can be obtained from Lockheed Missiles and Space Company, 3251 Hanover Street, Palo Alto, California.

9. SLADE PROGRAM [12]

SLADE is a special-purpose computer program for the static analysis of thin shells. It uses doubly curved, arbitrary quadrilateral element for mesh generation. Thin-shell theory is assumed to be valid for the analysis of shell structures. The considered types of shells are parts of the axisymmetric surfaces. SLADE can handle composite structures with a maximum of five layers and can also handle materials with a maximum of five elastic properties. Shells with variable wall thickness are solvable by this program. A computer with a 64,000-word core will be able to analyze a shell system with 650 quadrilateral elements and 700 nodal points. Mesh generation schemes are incorporated in the program to facilitate the data-preparation operations. SLADE's manual can be ordered from Sandia Laboratories, Livermore, California.

10. PROGRAM FEPLS [13]

Program FEPLS is a special purpose digital computer program for the finite element analysis of plane stress and plane strain problems. The idealization of the structure is done by constant-strain triangular elements.

The input information is the geometric configuration, loading conditions, boundary conditions, and material properties of the structural system. Once the input data is entered into the computer a general stiffness matrix is developed, then displacements at nodal points are determined. These displacement values are used for the calculation of nodal stresses. The user's manual for FEPLS can be ordered from the Bendix Corporation, Aerospace Systems Division, Ann Arbor, Mich.

A COMPARATIVE STUDY

It is neither practical nor possible to give the complete listings and documentations of all ten computer programs mentioned in the first part of this chapter. They are presented to familiarize the reader with some of the developed and available finite element computer programs. If a particular one is of a special interest, then it is suggested that the documentation and user manuals should be studied. Nevertheless, these ten programs present a significant overview of the possible practical applications of the finite element method.

The computer facilities at the University of Missouri-Rolla have a 360-50 digital computer with 260,000 bytes of core storage, six disk storage drives with 43×10^6 digits of random access storage. The computer programs which are operational at the center are: ICES-STRUDL-II; ELAS; "Z-C" Program; STRATA, and Wilson.

It is believed that a comparative evaluation between a few of the programs will have great engineering meaning. The author has used this approach in his graduate course on the finite element method and the results were well accepted by the students. The study presented here is an average outcome of a class project and is illustrative in nature. The reader should remember that all programs were run on IBM 360-50 systems.

Example 8-1

Perform the finite element analysis of the simply supported, tapered beam whose dimensions and loading condition are shown in Fig. 8-1. The

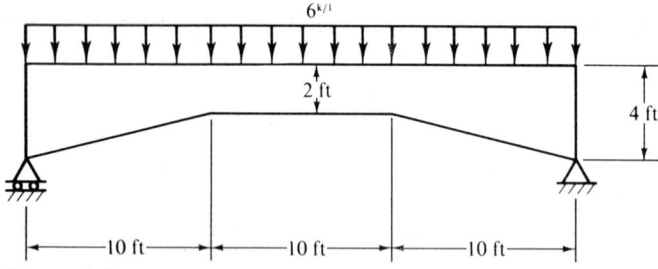

Figure 8-1

assumed Young's modulus is 30,000 ksi and the Poisson ratio is 0.3. Use the following computer programs to obtain the solution of the analysis.

    a Z-C Program
    b ICES-STRUDL-II Program
    c ELAS Program

Make a comparative study of the results from three points of view:

    a Computer time used (compilation and run times)
    b Input data preparation time
    c Compatibility of results

Solution

Since the given beam, Fig. 8-1, is symmetric, only one-half of the span is considered for the analysis. The solution is approximated by the plane stress analysis approach using constant-strain triangular elements.

The mesh used to obtain the results by the three computer programs is given in Fig. 8-2. The uniform load is used to simulate equivalent nodal point loads.

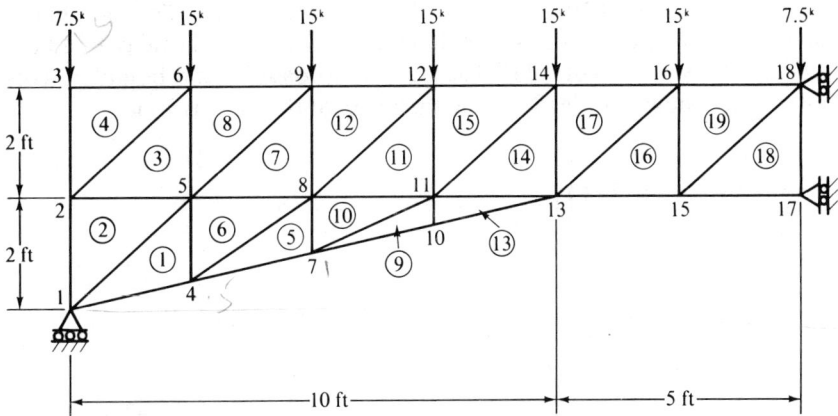

Figure 8-2

The results are shown in Table 8-2. The data preparation time is relative to the ability and knowledge of the individuals. The results are expected to be almost the same since the three programs are based on the same theory. The computer time might not be conclusive since the size of the problem and scheme of labeling the nodes might effect the efficiency.

TABLE 8-2

	Data Preparation Time, min	Computer Time Used, sec	Compatibility of Results	Type of Program
Z-C Program	30	35	All results were identical to third decimals	Special-purpose program
Ices-Strudl-II	25	34		General-purpose program
Elas	40	8		General-purpose program

Example 8-2

Perform the finite element analysis of the thin plate in plane stress by two computer programs:

   a ICES-STRUDL-II program
   b Z-C program

Then make a comparative study of the results obtained by the two programs, as well as the computer times used to perform the analysis. The plate is a composite one. The left half is made of steel (elements 1, 2, 5, 6) having a modulus of 30,000 ksi and Poisson ratio of 0.25. The right half is made of brass (elements 3, 4, 7, 8) having a modulus of elasticity of 15,000 ksi and Poisson ratio of 0.33. The geometry, loading, and boundary conditions are shown in Fig. 8-3. The thickness of the plate is 1 in.

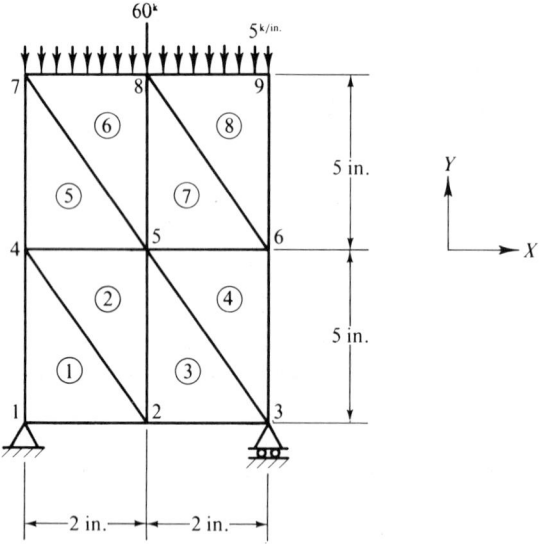

Figure 8-3

Solution
Constant-strain triangular elements are used for the idealization of the plate. Eight elements and nine nodal points define the topology of the system, Fig. 8-2. The comparison of the results are given in Table 8-3.

TABLE 8-3

Elements	σ_x		σ_y		γ_{xy}	
	STRUDL	Z-C	STRUDL	Z-C	STRUDL	Z-C
1	−1.7543	−1.7424	19.0355	19.1102	4.3858	4.3553
2	1.2173	1.2306	16.8968	17.0190	2.0049	2.0163
3	−0.2378	−0.1846	8.5577	8.6460	−5.7960	−5.9106
4	−0.3444	0.2338	15.5099	15.2244	−0.5946	−0.4616
5	0.7934	0.7555	15.6252	15.5938	−0.6407	−0.6097
6	−0.2562	−0.2440	23.1973	23.1259	6.2500	6.2375
7	−1.6667	−1.6398	11.4266	11.4003	1.7089	−1.6762
8	1.5602	1.5807	9.7510	9.8798	−3.9004	−3.9519

The computer usage time by the two programs to solve the same problem is given in Table 8-4.

TABLE 8-4

	STRUDL-II	"Z-C" Program
Computer run time (seconds)	22	97*

*Compilation time included.

Example 8-3

This is an illustrative solution of a thin flat plate in plane stress. The loading, geometry, and boundary conditions are shown in Fig. 8-4. The modulus of elasticity is 29,000 ksi, the Poisson ratio 0.27.

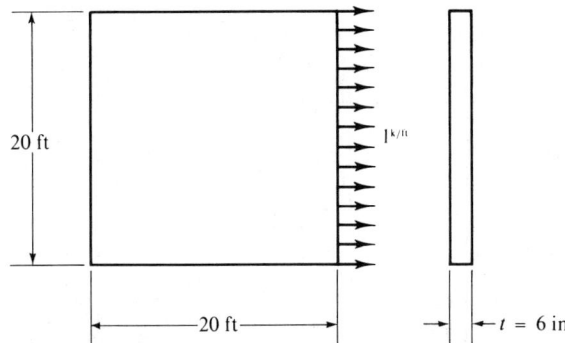

Figure 8-4

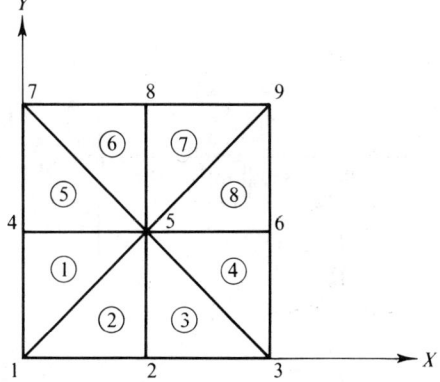

Figure 8-5

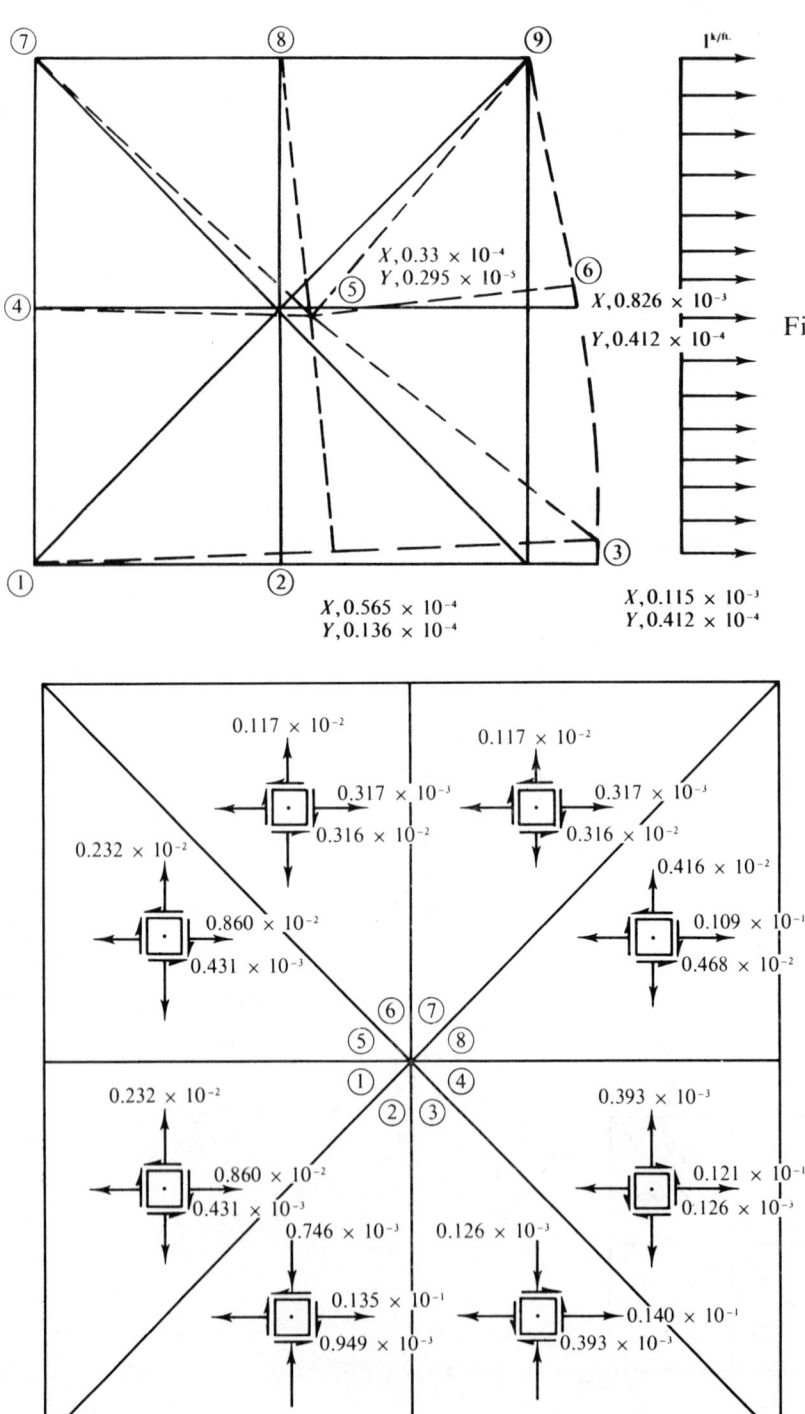

Figure 8-6

Figure 8-7 Stresses at the centroid of each element (in KSI).

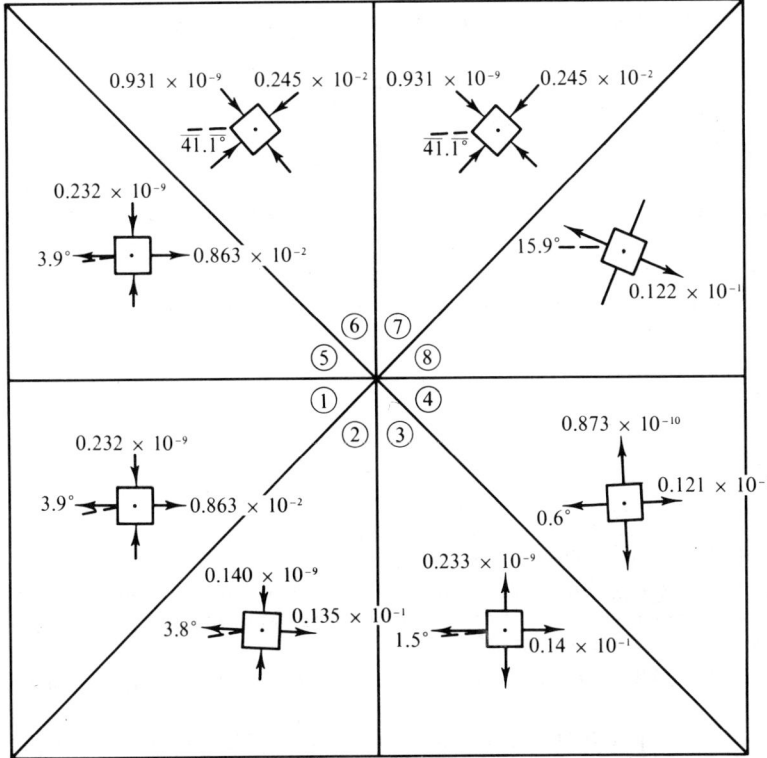

Figure 8-8 Principal stresses at the centroid of each element (in KSI).

Solution

Constant-strain triangular elements are used to generate the mesh as given in Fig. 8-5. The Z-C program is used for the analysis of this problem. The results obtained are nodal displacements, stresses at the centroids of the triangular elements, and finally the principal stresses with related principal angles.

The results are plotted for a quick engineering evaluation. Figure 8-6 gives the nodal displacements, Fig. 8-7 shows the stress plot, and Fig. 8-8 gives the principal stresses at the centroids.

PROBLEMS

8-1 Perform the analysis of the tapered beam discussed in Example 8-1. Prepare the input data for a computer program which is operational at the discretion of the reader. The results obtained can be compared with the STRUDL-II, ELAS, and Z-C outputs which are already given in the chapter.

8-2 Perform the finite element analysis of the Example 8-2 presented in the chapter. Prepare the input data for a computer program which is operational at your computer center. The results obtained can be compared with the STRUDL-II and Z-C outputs given in the chapter.

8-3 Perform a finite-element analysis of the Example 8-3 by any computer program and check the accuracy of the given results.

REFERENCES

1. R. D. LOGCHER, B. B. FLACHSBART, E. J. HALL, C. M. POWER, R. A. WELLS, JR., and A. J. FERRANTE, "ICES-STRUDL-II," Massachusetts Institute of Technology, Cambridge, Mass., 1968.
2. J. CONNOR and G. WILL, "Computer-Aided Teaching of the Finite Element Displacement Method," School of Engineering, Massachusetts Institute of Technology, Cambridge, Mass., 1969.
3. SP-222 NASTRAN User's Manual, Cosmic, University of Georgia, Athens, Ga., 1970.
4. SP-222 NASTRAN Theoretical Manual, Cosmic, University of Georgia, Athens, Ga., 1970.
5. R. M. JONES, and J. G. CROSE, "SAAS-II," Air Force Report No. SAMSO-TR-68-455, AD679 983, Clearninghouse, U. S. Department of Commerce, Springfield, Va., September 1968.
6. S. UTKU and F. A. AKYUZ, "ELAS—A General Purpose Computer Program for the Equilibrium Problems of Linear Structures," Vol. 1, User's Manual, National Aeronautics and Space Administration, Technical Report, 32-1240, Cosmic, University of Georgia, Athens, Ga.
7. R. L. DAVIS, and H. D. KEITH, "User's Manual for UMR Finite Element Computer Program—STRATA," Department of Engineering Mechanics, University of Missouri-Rolla, Rolla, Mo., 1969.
8. W. JOHNSON and P. B. MELLOR, *Plasticity for Mechanical Engineers*. New York: Van Nostrand, 1962, pp. 58–60.
9. O. C. ZIENKIEWICZ and Y. K. CHEUNG, *The Finite Element Method in Structural and Continuum Mechanics*. New York and London: McGraw-Hill, 1967.
10. E. L. WILSON, "Finite Element Analysis of Two-Dimensional Structures," Ph.D. thesis, University of California, Berkeley, Calif., 1963.
11. D. BUSHNELL, "Stress, Stability and Vibration of Complex Shells of Revolution: Analysis and User's Manual for BOSOR 3," Lockheed Missiles and Space Company, Sunnyvale, Calif., 1969.

12 S. W. KEY and Z. E. BEISINGER, "SLADE: A Computer Program for the Static Analysis of Thin Shells," SC-RR-69-369-Sandia Laboratories, Livermore, Calif., November 1970.

13 D. C. CHANG, "Plane Stress and Plane Strain Analysis by Finite Element Method," Program No. 0308, BSR 2907, The Bendix Corporation, Ann Arbor, Mich., August 1970.

The Finite Element Method and Ices-Strudl

INTRODUCTION

The majority of engineers are not inclined to write individual programs for the analysis of their problems; they prefer to make use of available computer programs which will efficiently solve their analysis problems. Their preference will be toward a programming system that is comparatively simple to learn and to use. The required input quantities should not demand sophisticated mathematical calculations. It is these needs of the engineers that stimulated interest among computer program developers to prepare general-purpose programs. The practical value of these user-oriented programs created one of the greatest engineering tools of our decade. ICES-STRUDL is one of the best known and widely used computer programming systems.

ICES PROGRAMS [1]

Integrated Civil Engineering Systems (ICES) is a user-oriented programming system including subsystems related to various specialty areas of civil engineering profession. It was developed by a team at the CIVIL ENGINEERing systems LABORATORY (CESL) of Massachusetts Institute of Technology. The list of the subsystems are given below:

    ICES-COGO   COordinate GEometry, dealing generally with surveying problems.
    ICES-SEPOL   SEttlement Problem Oriented Language.
    ICES-TRANSET   TRANsportation NETwork Analysis.
    ICES-ROADS   ROadway Analysis and Design System.
    ICES-BRIDGE   BRIDGE Design Systems.
    ICES-STRUDL   STRUctural Design Language.

Since our interests lie in the area of structural analysis using the finite-element method, our concern will be the understanding and use of the ICES-STRUDL subsystem. It might seem that this subsystem does not offer a direct academic contribution to the educational activities. ICES-STRUDL is valuable since it complements the whole concept of the finite element method by providing a quick and accurate solution to formulated analysis problems. It is obvious that the formulation of a problem does not end the necessary operations to obtain a solution. ICES-STRUDL will be a great help in performing the required matrix operations. *It is a dynamic program* that tends to change with added improved sections. These changes are minor in nature and can easily be learned. There are several versions of it available around the world, but they differ slightly. Therefore it is scientifically justifiable to present the basic input information of STRUDL in our discussion. The user of this program can be informed of the improvements through the periodic publications of his computer center.

The Structural Design Language [2] (STRUDL) is a structural information system containing both analysis and design capabilities. It is a general purpose program which is readily available in most of the computer centers.

STRUDL uses a command-oriented input approach. These commands are predetermined in a language familiar to the user. Each one will call for a computer operation related to the analysis. They can be grouped into five distinct categories:

    a Data input
    b Analytic procedures
    c Data subsetting
    d Data output
    e Output of analysis and design

The available analysis types which are operational are

- **a** Determinate analysis
- **b** Preliminary analysis
- **c** Stiffness analysis
- **d** Dynamic analysis
- **e** Finite element analysis
- **f** Nonlinear analysis
- **g** Linear buckling analysis

Additional capabilities of STRUDL include

- **a** Frame optimization
- **b** Steel design
- **c** Reinforced concrete design

The STRUDL subsystem has about two hundred commands to describe the input related to the analysis and design problems. The user describes the problem in detail to the computer using a set of proper commands. These set of commands form the input information. The commands are accepted by the computer and each command is processed individually and in the order in which they were read by the computer. The commands do specify well-defined operations: read the data, analyze the frame, write the results. The lack of logical order among the operations will stop the process. Hence, the engineer has the responsibility of knowing his problem well and having the ability to express it in well established logical steps.

We have stated that STRUDL is a user-oriented and command-communicated system. The data and instructions are conveyed to the computer by the proper commands. Hence the commands deserve serious consideration. The elements which make the commands are integers, which are numbers without decimal points. The symbols used as integers are $i_1, i_2, i_3, \ldots$; the symbols used as decimals are $V_1, V_2, V_3, \ldots$; alphanumerics which are combinations of numbers, letters, and blanks enclosed in single quotation marks. They are limited to eight characters, except for certain exceptions in titles. The symbols used for alphanumerics are $'a_1', 'a_2', 'a_3', \ldots$. The user is free to define the alphanumerics; words which are symbols, not defined by the user, are understood by the STRUDL. These names are not enclosed in single quotations as for alphanumerics.

ICES-STRUDL functions under the order of a set of well-defined commands, hence it is proper to list three commands to understand their use in the preparation of the input data. The detailed descriptions of these input preparation formats are available in the three volume publication of the Massachusetts Institute of Technology [3]. Here a concise description of all relevant commands related to STRUDL-II is presented. The instructor can enlarge this material if he feels his students need a more detailed discussion of the topic. However, the example problems at the end of this chapter, which are solved with STRUDL, will help to understand its use and data preparation for the analysis of structural systems. STRUDL can analyze one-, two-, and three-dimensional structures with ease and their basic inputs do not vary. The following itemized presentation has two goals.

    a The introduction of various commands related to STRUDL-II.
    b The logical order of commands for the preparation of the input data.

ICES-STRUDL subprogram was an extension of the STRESS [4] program and it had about twenty thousand source statements. The new version, known as STRUDL-II has more than forty thousand source statements. The information discussed in this chapter is related to STRUDL-II.

STRUDL-II COMMANDS

    1 Problem Initiation Command.
        STRUDL   'NAME'   'TITLE'
        'NAME' defines the problem using a maximum of eight alphanumeric characters.
        'TITLE' is the optional title of the problem using a maximum of sixty four characters.

Illustration
        STRUDL   'PROB 9'   'ILLUSTRATIVE PROCEDURE'

    2 Structural Analysis Type Command.
        This command will define the type of the problem to be solved by the program. The available types are:
            Plane truss
            Plane frame
            Plane grid
            Space truss

Space frame
Plane stress
Plane strain
Plate bending
Shallow shell

The last four types are more related to the finite element analysis and they will be used in the example problems formulated and solved in the second part of this chapter.

Illustration

    TYPE   PLANE   STRESS

3 Unit Command
The standard built-in units of the program are:
Length unit: inch
Force unit: pound
Angular unit: radian
Temperature unit: Fahrenheit
Time unit: second
If other units are desired, then they must be stated by a unit command.

Illustration

    UNIT   FEET   KG   MINUTE

4 Joint Coordinate Command
This command will define the geometry of the structure by giving the coordinates of the nodal points. The first column is the nodal points. The second column is the X coordinate, the third column the Y coordinate, and the fourth column the Z coordinate. The S in the fifth column defines the support condition of any type. The details of the support will be defined by another command, namely, the *joint release command.*

Illustration

1	X	17.0	Y	7.0	Z	0.0	S
2		10.0		1.0		0.0	
3		5.0		3.0		0.0	

5 Member Incidences Command
This command defines the topology of the structural system by relating the elements with their nodes.

Illustration

ELEMENT	INCIDENCES		
1	4	5	6
2	9	10	15

6. Surface Equation Command
   This command defines the surface related to the problem to be analyzed by giving the proper curvatures. This command is associated with the shell analysis.

Illustration

SURFACE EQUATION

'HYSHELL'   XY   .007

7. Element Properties Command
   This command defines the element properties.

Illustration

ELEMENT   PROPERTIES

10 TO 17   TYPE   'PSR'   THICKNESS   1.6

8. Element Constants Command
   This command specifies the physical properties of the elements as well as the element orientation with respect to global coordinates system. The relevant constants are Young's modulus, E; shear modulus, G; density of member material, DEN; coefficient of thermal expansion, CTE; orientation angle, BETA.

Illustration

CONSTANTS

E   30000.   ALL   BUT   29000.   2

POISSON   0.30   ALL

9. Joint Release Command
   In the joint coordinates command an S was used to define any type of support condition. To make the boundary conditions be compatible with the actual supports, proper releases are necessary.

Illustration

JOINT RELEASES

2   FORCE X   MOMENT Y

10 Loading Command
This command specifies the loading identification, types, directions and magnitudes. The loading can be uniform, concentrated and linear as well as joint loads and surface loads.

JOINT LOADS

Joint Name   Force   its   direction   its   magnitude

Illustration

1   FORCE   X   10.0   Y   20.0

SURFACE LOADS

Element Name   Surface Forces Global PZ

Illustration

ELEMENT LOADS

5 to 14   SURFACE FORCES GLOBAL PZ 1.6

The loads in the plane stress, plane strain and plate bending problems are given in global coordinates; the loadings for shell problems are given in local coordinates of the elements.

11 Analysis Command
All data of the problem must precede this particular command. The finite element analysis uses stiffness matrix analysis, therefore this command will be STIFFNESS ANALYSIS.

12 Output Command
This command will print the results of the analysis. In the case of shell problems, the results are given at the centroid of the elements; in the case of other problems, the results are given at the nodes. The command has the following form:

LIST   DISPLACEMENTS,   STRESSES,   STRAINS,   REACTIONS.

Illustration

LIST   DISPLACEMENTS   JOINTS   1 TO 12

LIST   STRESSES,   STRAINS   ALL

13 Plot Command
This command will produce graphical output for structural geometry and analysis of the problem.

14 Termination Command
This command ends the analysis process, and it has the following form:

FINISH.

This ends the discussion of the input and output commands necessary for the formulation of an analysis problem.

STRUDL-II AND THE FINITE ELEMENT METHOD

ICES-STRUDL II is able to perform linear finite element analyses of various structural systems [5]. The sequence of operations are as follows:

Input
Compilation and Checking
Element Stiffness Matrix Generation
Assemble Global Stiffness Matrix
Joint Load Processor
Solution of Node Displacements
Solution for Nodal Strains and Stresses
Output

Among these eight phases, the ones that are element-dependent are the Element Stiffness Matrix Generation, Computations of Nodal Strains, and Stresses and Output. To increase the efficiency of the program, it is suggested that the nodes should be numbered in "short directions" of the geometry of the structure. This procedure will secure minimum bandwidth, making the matrix storage more economical and matrix operations more efficient.

ICES-STRUDL II can solve three types of problems in structural analysis using the finite element technique:

    a Plane stress and plane strain problems
    b Flat plate in bending problems
    c Shallow shell-bending problems

The STRUDL-II subsystem has built-in element-types that correspond to a particular type of problem. These elements are listed below.

    a Plane Stress and Plane Strain Type Elements:
        Linear Strain Triangle (LST)
        Constant Strain Triangle-Global formulation (CSTG)
        Constant Strain Triangle-Local formulation (CSTL)
        Plane Stress/Strain Rectangle (PSR)

For a particular plane problem any combination of the above elements can be used. The choice of the element type will be decided by the geometric shape of the structure to be analyzed.

b Plate Bending Elements:
Flat Plate Triangle (CPT)
Flat Plate Rectangle (BPR)
Flat Plate Parallelogram (BPP)

Anyone or a combination of the above elements can be used for the solution of a plate problem.

c Shallow Shell Element:
Shallow Shell Curved Rectangle (SSCR)

Finite Element Analysis with STRUDL

We have discussed theoretically how STRUDL is applied to the analysis of structural problems using the finite element method. It is believed that the understanding of STRUDL's application will be more complete if a few problems are formulated for it and solved by it. The mesh generation for these examples are simple with comparatively few elements since they tend to be illustrative. If more accurate results are desired, the idealization of the structural system must be refined. Each problem's input data is printed by the computer. The reader should compare them with the commands presented in the first half of this chapter. The output for each problem is also printed. Therefore each individual example presents a complete STRUDL solution. The reader should be aware of the fact that if the element stiffness matrices are generated in local coordinates, they need to be transformed to the same global coordinate system before they are superimposed. The only mean to avoid this coordinate transformation is to generate all element stiffness matrices k_{ij} in the same coordinate system, which is rather difficult. Even though STRUDL will perform all the necessary computations internally, the results obtained as output must be carefully interpreted to avoid engineering errors.

Example 9-1

Perform the analysis of the simply supported elastic beam shown in Fig. 9-1, using the displacement finite element method. Prepare the input data necessary for STRUDL and use this general purpose program to obtain the solution of the problem. The beam is loaded by a uniform load of 2 kips/ft. The Young's modulus is 30,000 ksi and the Poisson ratio is 0.3. The geometry of the beam is given in the Fig. 9-1.

214 THE FINITE ELEMENT METHOD AND ICES-STRUDL

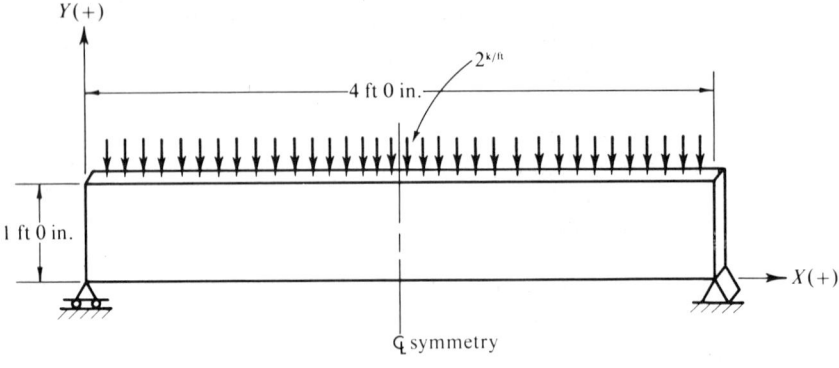

$E = 30{,}000$ ksi $\mu = 0.3$ $t = 1$ in.

Figure 9-1

Solution

Since the beam is symmetric from the points of view of geometry and loading, only one-half of the span needs to be considered for analysis. The mesh generation uses only eight triangular elements in the half of the structure for the simplicity is shown in Fig. 9-2. The beam problem is analyzed as a plane stress problem using constant strain triangular elements. This type of finite element will provide a good approximation if a fine mesh is developed. The same problem was analyzed by longhand operation in Chapter 5. STRUDL input quantities will be now prepared

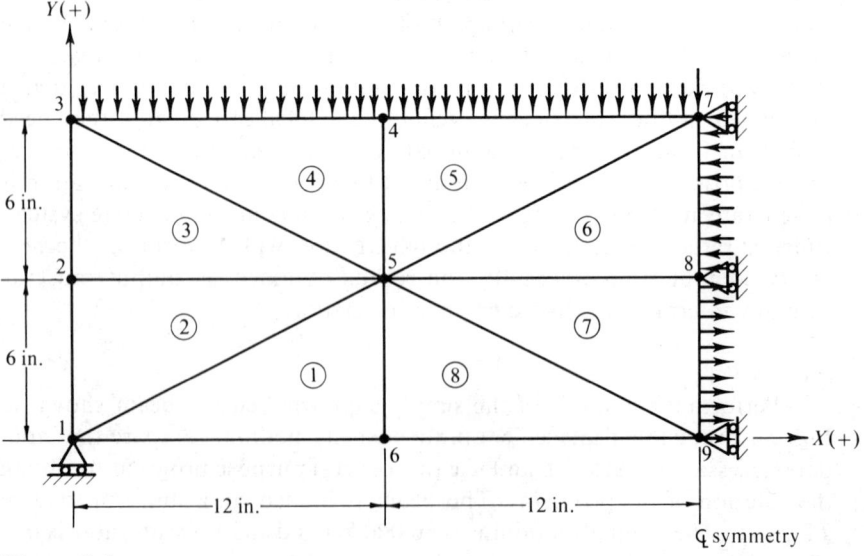

Figure 9-2

according to the commands necessary for this problem described in the first part of Chapter 9. To obtain the solution, a digital computer is used for practical purposes.

```
STRUDL INPUT:
STRUDL    'PROB 9-1'    'ANALYSIS OF A SIMPLY
SUPPORTED BEAM'
TYPE    PLANE STRESS
UNITS KIPS INCHES
JOINT COORDINATES
1    0.    0.    S
2    0.    6.
3    0.    12.
4    12.   12.
5    12.   6.
6    12.   0.
7    24.   12.   S
8    24.   6.    S
9    24.   0.    S

ELEMENT INCIDENCES
1    1    6    5
2    1    5    2
3    2    5    3
4    3    5    4
5    4    5    7
6    5    8    7
7    5    9    8
8    5    6    9

ELEMENT PROPERTIES
1 TO 8 TYPE 'CSTG' THICKNESS 1.
$ 'CSTG' IS A CONSTANT STRAIN TRIANGULAR ELEMENT
CONSTANTS
E 30000. ALL
POISSON 0.3 ALL
$ RECOGNIZE SYMMETRY IN SPECIFIC RELEASES
JOINT RELEASES
7 FORCE Y
8 FORCE Y
9 FORCE Y
LOADING 'ONE'
JOINT LOADS
JOINT 3 LOAD FORCE Y −1.
JOINT 4 LOAD FORCE Y −2.
JOINT 7 LOAD FORCE Y −1.
LOADING LIST ALL
STIFFNESS ANALYSIS
PRINT DATA
LIST STRESSES, STRAINS, FORCES, DISPLACEMENTS
FINISH
```

TABLE 9-1 PROBLEM 9-1. ANALYSIS OF A SIMPLY SUPPORTED BEAM.

```
ACTIVE UNITS     INCH KIP  RAD  DEGF SEC
ACTIVE STRUCTURE TYPE   PLANE   STRESS
ACTIVE COORDINATE AXES  X  Y

LOADING - ONE

ELEMENT STRESSES
```

ELEMENT						
1	SXX	-0.258779E 00	SYY	-0.163241E 00	SXY	-0.303428E 00
2	SXX	-0.233493E 00	SYY	-0.514955E 00	SXY	-0.316948E 00
3	SXX	-0.162725E 00	SYY	-0.279062E 00	SXY	-0.316840E 00
4	SXX	-0.273679E 00	SYY	-0.450107E-01	SXY	-0.102066E 01
5	SXX	-0.441113E 00	SYY	-0.156141E 00	SXY	-0.395737E-01
6	SXX	-0.276883E 00	SYY	-0.186454E 00	SXY	-0.222891E 00
7	SXX	-0.210016E 00	SYY	0.364356E-01	SXY	0.107791E 01
8	SXX	0.202335E 00	SYY	-0.249066E-01	SXY	-0.728717E-01

ELEMENT PRINCIPAL STRESSES

ELEMENT						
1	S1	0.961553E-01	S2	-0.518175E 00	TMAX	0.307165E 00
2	S1	-0.109122E-01	S2	-0.737535E 00	TMAX	0.363311E 00
3	S1	0.122035E-01	S2	-0.369583E 00	TMAX	0.148690E 00
4	S1	-0.928510E-01	S2	-0.411540E 00	TMAX	0.252196E 00
5	S1	-0.423792E-02	S2	-0.459099E 00	TMAX	0.247175E 00
6	S1	-0.467896E 00	S2	-0.241476E 00	TMAX	0.254686E 00
7	S1	0.236895E 00	S2	-0.462672E-01	TMAX	0.134981E 00
8						

ELEMENT STRAINS

ELEMENT						
1	EXX	-0.699357E-05	EYY	-0.285357E-05	EXY	-0.262971E-04
2	EXX	-0.263354E-05	EYY	-0.148303E-04	EXY	-0.290288E-04
3	EXX	-0.102335E-05	EYY	-0.767481E-05	EXY	-0.118594E-04
4	EXX	-0.867220E-05	EYY	0.123643E-05	EXY	0.194818E-04
5	EXX	-0.199090E-04	EYY	-0.344631E-05	EXY	-0.342973E-05
6	EXX	-0.736489E-05	EYY	-0.285357E-05	EXY	-0.193172E-04
7	EXX	-0.736489E-05	EYY	-0.331468E-05	EXY	-0.193172E-04
8	EXX	-0.699357E-05	EYY	-0.285357E-05	EXY	-0.631556E-05

LOADING - ONE

MEMBER   FORCES

MEMBER   JOINT   AXIAL FORCE

RESULTANT JOINT DISPLACEMENTS - SUPPORTS

JOINT	/----------------DISPLACEMENT----------------/		
	X DISP.	Y DISP.	Z DISP.
1	0.0	0.0009298	
7	0.0	-0.0009091	
8	0.0		
9	0.0	-0.0009290	

RESULTANT JOINT DISPLACEMENTS - FREE JOINTS

JOINT	/----------------DISPLACEMENT----------------/		
	X DISP.	Y DISP.	Z DISP.
2	0.0001200	-0.0000890	
3	0.0003430	-0.0001350	
4	0.0002389	-0.0006699	
5	0.0000884	-0.0006773	
6	-0.0000839	-0.0006602	

STRUDL OUTPUT:

The computer output of the analysis of the simply supported beam is attached as Table 9-1. There are four sets of output.

- a Element stresses
- b Element strains
- c Resultant joint loads
- d Resultant joint displacements

Example 9-2

Perform the analysis of the thin plate shown in Fig. 9-3, which is under pure tension due to the nature of the applied loads, using the finite element method and STRUDL program. Young's modulus is 30,000 ksi and the Poisson ratio is 0.3. The plate is supported along one edge. Its geometric dimensions are given in Fig. 9-3. Its thickness is 2 in.

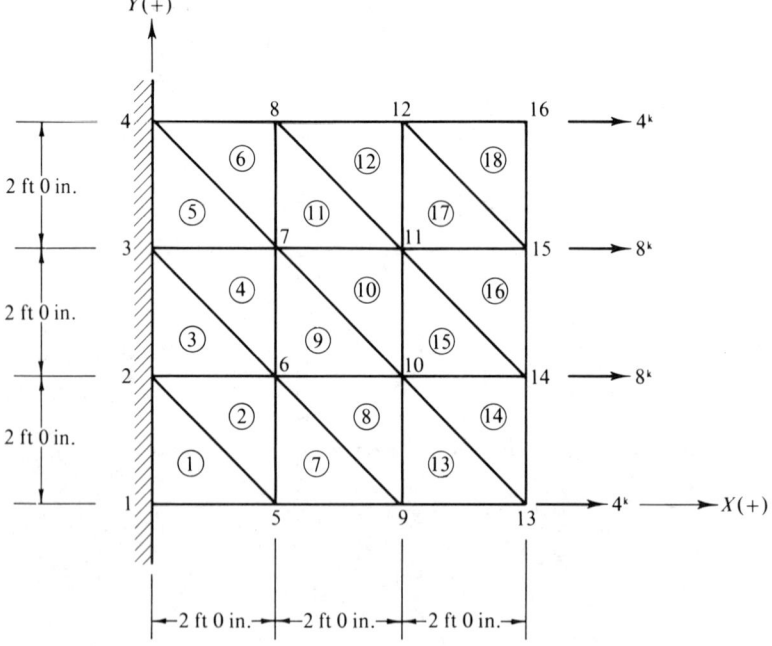

Figure 9-3

Solution

The thin plate is divided into 18 triangular elements. It is supported at the nodes 1, 2, 3, and 4. The tensile loads are applied at nodes 13, 14, 15 and 16 as shown in Fig. 9-3. Since this is a plane stress problem, the elements used are constant-strain triangles. The solution of this problem is obtained using the STRUDL program. The input data is prepared according to the commands defined in the first part of this chapter.

STRUDL INPUT:
STRUDL 'PROB 9-2'   'THIN PLATE IN PLANE STRESS ANALYSIS'
$ CONSTANT STRAIN TRIANGULAR ELEMENTS ARE USED
$ FOR THE ANALYSIS OF THE PLATE.
TYPE   PLANE STRESS
UNITS KIPS FEET
JOINT COORDINATES
 1   0.   0.   S
 2   0.   2.   S
 3   0.   4.   S
 4   0.   6.   S
 5   2.   0.
 6   2.   2.
 7   2.   4.
 8   2.   6.
 9   4.   0.
10   4.   2.
11   4.   4.
12   4.   6.
13   6.   0.
14   6.   2.
15   6.   4.
16   6.   6.

ELEMENT INCIDENCES
 1   1   5   2
 2   2   5   6
 3   2   6   3
 4   3   6   7
 5   3   7   4
 6   4   7   8
 7   5   9   6
 8   6   9   10
 9   6   10   7
10   7   10   11
11   7   11   8
12   8   11   12
13   9   13   10
14   10   13   14
15   10   14   11
16   11   14   15
17   11   15   12
18   12   15   16

UNITS INCHES
ELEMENT PROPERTIES
1 TO 18 TYPE 'CSTG' THICKNESS 2.0

TABLE 9-2 PROBLEM 9-2. THIN PLATE ANALYSIS BY F.E.M.

ACTIVE UNITS INCH KIP RAD DFGF.SEC
ACTIVE STRUCTURE TYPE PLANE STRESS
ACTIVE COORDINATE AXES X Y

LOADING - ONE   POINT

ELEMENT STRESSES

ELEMENT	SXX		SYY		SXY	
1	SXX	0.169898E 00	SYY	0.509994E-03	SXY	0.204365E-01
2	SXX	0.154343E 00	SYY	0.100100E-02	SXY	0.569248E-02
3	SXX	0.167770E 00	SYY	0.500920E-03	SXY	-.166730E-02
4	SXX	0.159293E 00	SYY	0.512243E-02	SXY	-.728840E-02
5	SXX	0.172257E 00	SYY	0.107241E-01	SXY	-.187505E-02
6	SXX	0.174146E 00	SYY	0.106809E-01	SXY	-.183875E-02
7	SXX	0.163332E 00	SYY	0.138560E-02	SXY	-.148217E-02
8	SXX	0.188364E 00	SYY	-.235142E-02	SXY	-.482173E-02
9	SXX	0.180140E 00	SYY	0.645280E-02	SXY	-.208322E-02
10	SXX	0.164430E 00	SYY	0.454188E-02	SXY	-.294862E-02
11	SXX	0.164333E 00	SYY	0.116689E-02	SXY	-.208222E-02
12	SXX	0.165148E 00	SYY	0.164338E-02	SXY	-.116784E-02
13	SXX	0.166089E 00	SYY	0.116050E-02	SXY	-.153105E-02
14	SXX	0.167438E 00	SYY	0.119080E-02	SXY	-.348105E-02
15	SXX	0.166814E 00	SYY	0.192050E-02	SXY	-.914713E-03
16	SXX	0.167190E 00	SYY	0.156973E-02	SXY	-.230454E-02
17	SXX	0.165710E 00	SYY	-.956680E-03	SXY	-.956741E-03

ELEMENT PRINCIPAL STRESSES

ELEMENT	S1		S2		TMAX		THETA	
1	S1	0.173312E 00	S2	0.173412E-01	TMAX	0.475556E-01	THETA	0.628780E-01
2	S1	0.154278E 00	S2	0.000000E 00	TMAX	0.775595E-01	THETA	0.732563E-01
3	S1	0.167047E 00	S2	0.500300E-03	TMAX	0.497540E-01	THETA	0.586466E-01
4	S1	0.159563E 00	S2	0.300000E 00	TMAX	0.122230E-01	THETA	0.734202E-01
5	S1	0.180777E 00	S2	0.000000E 00	TMAX	0.507368E-02	THETA	0.602600E-01
6	S1	0.186511E 00	S2	0.000000E 00	TMAX	0.756380E-02	THETA	0.866610E-01
7	S1	0.164333E 00	S2	0.000000E 00	TMAX	0.106696E-01	THETA	0.764076E-01
8	S1	0.188666E 00	S2	0.000000E 00	TMAX	0.133128E-01	THETA	0.814849E-01
9	S1	0.188953E 00	S2	0.000000E 00	TMAX	0.158156E-01	THETA	0.764322E-01
10	S1	0.188953E 00	S2	0.000000E 00	TMAX	0.230706E-02	THETA	0.833229E-01
11	S1	0.170167E 00	S2	0.000000E 00	TMAX	0.646490E-02	THETA	0.818513E-01
12	S1	0.164480E 00	S2	0.000000E 00	TMAX	0.448696E-02	THETA	0.820172E-01
13	S1	0.165141E 00	S2	0.000000E 00	TMAX	0.159681E-02	THETA	0.838855E-01
14	S1	0.166528E 00	S2	0.000000E 00	TMAX	0.114876E-02	THETA	0.826456E-01
15	S1	0.167439E 00	S2	0.000000E 00	TMAX	0.191981E-02	THETA	0.764785E-01
16	S1	0.167020E 00	S2	0.000000E 00	TMAX	0.231403E-02	THETA	0.844684E-01
17	S1	0.182828E 00	S2	0.000000E 00	TMAX	0.153786E-02	THETA	0.834180E-01
18	S1	0.165746E 00	S2	0.000000E 00	TMAX	-.962198E-03	THETA	0.833389E-01

ELEMENT STRAINS

ELEMENT						
1	EXX	0.515357E-05	EYY	-0.127766E-05	EXY	0.177116E-05
2	EXX	0.506870E-05	EYY	-0.115984E-05	EXY	0.398627E-06
3	EXX	0.517985E-05	EYY	-0.148503E-05	EXY	-0.493507E-06
4	EXX	0.587946E-05	EYY	-0.127766E-05	EXY	-0.545177E-06
5	EXX	0.544247E-05	EYY	-0.159783E-05	EXY	-0.666335E-06
6	EXX	0.564408E-05	EYY	-0.115984E-05	EXY	-0.156759E-07
7	EXX	0.566644E-05	EYY	-0.161037E-05	EXY	-0.526621E-07
8	EXX	0.547790E-05	EYY	-0.148503E-05	EXY	-0.118056E-06
9	EXX	0.548838E-05	EYY	-0.159783E-05	EXY	-0.255552E-06
10	EXX	0.562208E-05	EYY	-0.170384E-05	EXY	-0.183546E-06
11	EXX	0.559063E-05	EYY	-0.161037E-05	EXY	-0.925441E-07
12	EXX	0.553324E-05	EYY	-0.174857E-05	EXY	-0.100060E-06
13	EXX	0.553324E-05	EYY	-0.162657E-05	EXY	-0.132916E-06
14	EXX	0.553324E-05	EYY	-0.168899E-05	EXY	-0.301691E-07
15	EXX	0.553324E-05	EYY	-0.168899E-05	EXY	-0.792751E-08
16	EXX	0.553324E-05	EYY	-0.168899E-05	EXY	0.199725E-06
17	EXX	0.553324E-05	EYY	-0.168899E-05	EXY	0.199725E-06
18	EXX	0.553324E-05	EYY	-0.168899E-05	EXY	0.629176E-07

LOADING - ONE   POINT

RESULTANT JOINT DISPLACEMENTS - SUPPORTS

JOINT	/------------DISPLACEMENT------------/		
	X DISP.	Y DISP.	Z DISP.
1	0.0	0.0	0.0
2	0.0	0.0	0.0
3	0.0	0.0	0.0
4	0.0	0.0	0.0

RESULTANT JOINT DISPLACEMENTS - FREE JOINTS

JOINT	/------------DISPLACEMENT------------/		
	X DISP.	Y DISP.	Z DISP.
5	0.0001237	0.0000425	
6	0.0001214	0.0000118	
7	0.0001243	-0.0000160	
8	0.0001411	-0.0000516	
9	0.0002515	0.0000482	
10	0.0002529	0.0000102	
11	0.0002726	-0.0000284	
12	0.0002836	-0.0000676	
13	0.0003836	0.0000455	
14	0.0003866	0.0000046	
15	0.0003930	-0.0000374	
16	0.0004054	-0.0000779	

```
CONSTANTS
E 30000. ALL
POISSON 0.3 ALL
LOADING 'ONE' 'POINT'
JOINT LOADS
13 FORCE X 4.0
14 FORCE X 8.0
15 FORCE X 8.0
16 FORCE X 4.0
LOADING LIST ALL
STIFFNESS ANALYSIS
PRINT DATA
LIST STRESSES, STRAINS, FORCES, DISPLACEMENTS
FINISH
```

STRUDL OUTPUT:
The results of the plate analysis are given in four groups.

a Element stresses
b Element strains
c Nodal forces
d Nodal displacements

The computer output is given in Table 9-2.

Example 9-3

Perform the finite-element analysis of the plate in bending shown in Fig. 9-4. It is supported along two adjacent edges, as fixed against translation and rotation. There are two types of loading: first it has a uniform load of 250 lb/sq ft of surface area and a concentrated load of 2 kips at the free corner. Young's modulus is 30,000 ksi and the Poisson ratio is

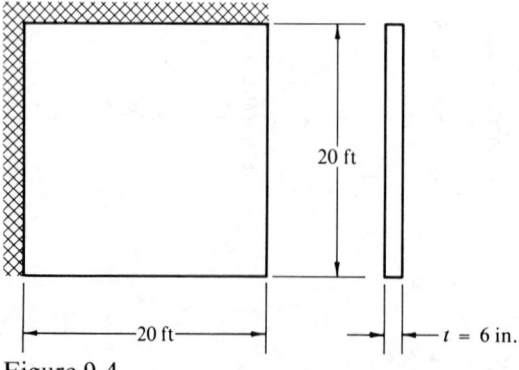

Figure 9-4

0.3. The geometric dimensions of the thin plate are given in Fig. 9-4. The thickness of the plate is 6 in.

Solution

Since this is an illustrative example, only eight flat-plate triangular elements are used in the mesh generation. The analysis of a plate in bending follows the theory discussed in Chapter 7. The mesh generation and topology is shown in Fig. 9-5. The plate is supported at nodes 1, 4, 7, 8,

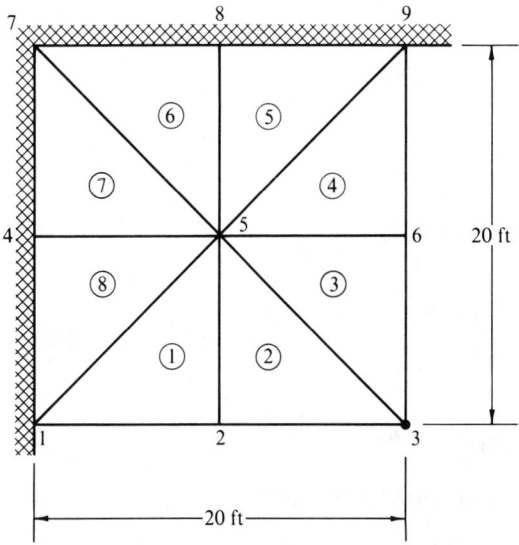

t = 6 in. Loading: 1 Uniform on ℞ 250 lb/ft²
ν = 0.3 2 Concentrated @ node ③ = 2ᵏ
E = 30,000 ksi

Figure 9-5

and 9. It is loaded perpendicular to its plane, causing a bending analysis condition. The solution is obtained using STRUDL and STRUDL input is prepared according to the commands described in the first part of this chapter.

```
STRUDL INPUT
STRUDL 'PROB 9-3'    'ANALYSIS OF A PLATE IN BENDING'
$ EIGHT FLAT PLATE TRIANGULAR ELEMENTS ARE USED.
TYPE BENDING
UNIT FEET
JOINT COORDINATES
  1    0.0     0.0   S
  2   10.0     0.0
  3   20.0     0.0
```

```
4      0.0    10.0   S
5     10.0    10.0
6     20.0    10.0
7      0.0    20.0   S
8     10.0    20.0   S
9     20.0    20.0   S
ELEMENT INCIDENCES
1    1   2   5
2    2   3   5
3    3   6   5
4    6   9   5
5    9   8   5
6    5   8   7
7    5   7   4
8    5   4   1
```
UNITS INCHES KIPS
ELEMENT PROPERTIES
1 TO 8 TYPE 'CPT' THICKNESS 6.0
CONSTANTS
E 30000 ALL
POISSON 0.3 ALL
UNIT FEET
LOADING 'ONE' 'POINT'
JOINT LOADS
3 FORCE Z −2.0
LOADING 'TWO' 'UNIFORM'
ELEMENT LOADS
1 TO 8 SURFACE FORCES GLOBAL PZ −0.25
LOADING LIST ALL
STIFFNESS ANALYSIS
LIST DISPLACEMENTS, STRAINS, STRESSES, REACTIONS, ALL
FINISH

STRUDL OUTPUT:

The results are divided into two parts relative to the concentrated and uniform loadings. Their actions are analyzed independently without combination. If desired, they can be automatically superimposed. Each output has four groups of information.

a Element stresses
b Element strains
c Nodal forces
d Nodal displacements

The computer output is given in Table 9-3.

TABLE 9-3 PROBLEM 9-3. ANALYSIS OF A PLATE IN BENDING.

```
ACTIVE UNITS  FEET KIP RAD DEGF SEC
ACTIVE STRUCTURE TYPE PLATE BENDING
ACTIVE COORDINATE AXES X Y

LOADING - ONE         POINT

ELEMENT STRESSES

ELEMENT
 1     NODE  5    MXX  0.114721E 01   MYY  0.598269E 00   MXY -0.481355E 00
 1     NODE  2    MXX  0.181633E 01   MYY  0.126034E-01   MXY -0.260893E 00
 1     NODE  3    MXX -0.378288E 00   MYY -0.378290E 00   MXY -0.371124E 00
 2     NODE  5    MXX  0.369008E 00   MYY  0.363007E-01   MXY -0.784066E 00
 2     NODE  3    MXX  0.739107E 00   MYY  0.363005E 00   MXY -0.481356E 00
 2     NODE  6    MXX -0.378289E 00   MYY -0.378210E 00   MXY -0.641356E 00
 3     NODE  6    MXX -0.739121E 00   MYY -0.739211E 00   MXY -0.784066E 00
 3     NODE  9    MXX  0.126229E-01   MYY  0.181463E-01   MXY -0.608881E 00
 3     NODE  6    MXX -0.598263E-01   MYY -0.147201E 01   MXY  0.260894E 00
 4     NODE  9    MXX -0.322991E-01   MYY -0.179331E 01   MXY -0.481355E 00
 4     NODE  6    MXX -0.228565E 00   MYY -0.320372E 00   MXY -0.371125E 00
 5     NODE  8    MXX -0.235200E 00   MYY  0.494198E 00   MXY -0.188895E-00
 5     NODE  9    MXX -0.301976E-01   MYY  0.494023E 00   MXY -0.188895E-06
 5     NODE  6    MXX -0.737144E-06   MYY  0.889475E-06   MXY -0.447150E-06
 6     NODE  5    MXX -0.555120E 00   MYY  0.555630E 00   MXY -0.244672E 06
 6     NODE  8    MXX  0.244217E 00   MYY  0.555630E 00   MXY -0.137236E 03
 6     NODE  6    MXX  0.555629E-07   MYY  0.556306E-06   MXY -0.137236E 03
 7     NODE  5    MXX  0.574554E-06   MYY  0.555630E-06   MXY -0.495849E-06
 7     NODE  8    MXX  0.219555E-06   MYY  0.244272E 00   MXY -0.137236E-05
 7     NODE  4    MXX  0.784198E-06   MYY  0.244272E 00   MXY -0.137235E-06
 8     NODE  5    MXX  0.320373E 00   MYY -0.235259E 00   MXY -0.425013E-06
 8     NODE  4    MXX  0.494023E 00   MYY -0.228566E 00   MXY -0.274469E 00
 8     NODE  4    MXX  0.494023E 00   MYY -0.301979E-01   MXY -0.137234E 00
```

ELEMENT STRAINS

ELEMENT

LOADING - TWO   UNIFORM

TABLE 9-3 (continued).

ELEMENT STRESSES

ELEMENT						
1	NODE 5	MXX	0.642475E 01	MYY	0.279094E 00	MXY -0.270670E 01
1	NODE 1	MXX	0.263226E 02	MYY	0.201769E 02	MXY 0.156648E 01
2	NODE 3	MXX	-0.276194E 01	MYY	-0.269786E 01	MXY 0.318087E 01
2	NODE 5	MXX	-0.235477E 01	MYY	-0.235479E 01	MXY -0.293787E 00
2	NODE 2	MXX	-0.188351E 01	MYY	-0.257695E 01	MXY 0.270672E 00
3	NODE 5	MXX	-0.276197E 01	MYY	-0.276198E 01	MXY -0.293800E 00
3	NODE 6	MXX	-0.257697E 01	MYY	-0.188354E 01	MXY -0.154650E 01
4	NODE 9	MXX	-0.201769E 02	MYY	-0.263226E 02	MXY -0.156648E 01
4	NODE 5	MXX	-0.279743E 00	MYY	-0.642471E 01	MXY 0.270671E 01
5	NODE 6	MXX	-0.369782E 00	MYY	-0.710619E 00	MXY -0.307283E 00
5	NODE 9	MXX	-0.101600E 02	MYY	-0.401432E 01	MXY -0.188895E 05
6	NODE 8	MXX	-0.263386E 01	MYY	-0.877952E 00	MXY 0.528642E-05
6	NODE 5	MXX	-0.169894E-04	MYY	0.113778E-02	MXY 0.369786E-05
6	NODE 8	MXX	-0.138048E 01	MYY	-0.150308E-02	MXY 0.307283E 01
7	NODE 5	MXX	-0.138050E 01	MYY	0.180048E 01	MXY 0.263645E 01
7	NODE 7	MXX	-0.750864E-05	MYY	0.305000E-01	MXY -0.307284E 01
8	NODE 4	MXX	0.830502E-01	MYY	0.138049E-04	MXY 0.500116E-05
8	NODE 1	MXX	0.877952E 01	MYY	0.448884E 01	MXY 0.136642E-05
8	NODE 5	MXX	0.401431E 01	MYY	0.263185E 02	MXY -0.434458E-05
8	NODE 2	MXX	-0.113778E-02	MYY	0.141699E 01	MXY -0.307283E 01
						-0.153641E 01

ELEMENT STRAINS

ELEMENT

TABLE 9-3 (continued).

LOADING - ONE   POINT

RESULTANT JOINT LOADS - SUPPORTS

JOINT	X FORCE	Y FORCE	Z FORCE	X MOMENT	Y MOMENT	Z MOMENT
1			0.0	0.0	0.0	
4			0.0	0.0	0.0	
7			0.0	0.0	0.0	
8			0.0	0.0	0.0	
9			0.0	0.0	0.0	

RESULTANT JOINT DISPLACEMENTS - SUPPORTS

JOINT	X DISP.	Y DISP.	Z DISP.	X ROT.	Y ROT.	Z ROT.
1			0.0	0.0	0.0	
4			0.0	0.0	0.0	
7			0.0	0.0	0.0	
8			0.0	0.0	0.0	
9			0.0	0.0	0.0	

RESULTANT JOINT DISPLACEMENTS - FREE JOINTS

JOINT	X DISP.	Y DISP.	Z DISP.	X ROT.	Y ROT.	Z ROT.
2			-0.0012585	0.0000754	0.0002183	
3			-0.0039897	0.0002858	0.0002858	
5			-0.0003846	0.0000793	0.0000793	
6			-0.0012585	0.0002183	0.0000754	

LOADING - TWO   UNIFORM

RESULTANT JOINT LOADS - SUPPORTS

JOINT	X FORCE	Y FORCE	Z FORCE	X MOMENT	Y MOMENT	Z MOMENT
1			0.0	0.0	0.0	
4			0.0	0.0	0.0	
7			0.0	0.0	0.0	
8			0.0	0.0	0.0	
9			0.0	0.0	0.0	

227

TABLE 9-3 (continued).

RESULTANT JOINT DISPLACEMENTS - SUPPORTS

JOINT	/------DISPLACEMENT------//			/------ROTATION------/		
	X DISP.	Y DISP.	Z DISP.	X ROT.	Y ROT.	Z ROT.
1	0.0	0.0	0.0	0.0	0.0	0.0
4	0.0	0.0	0.0	0.0	0.0	0.0
7	0.0	0.0	0.0	0.0	0.0	0.0
8	0.0	0.0	0.0	0.0	0.0	0.0
9	0.0	0.0	0.0	0.0	0.0	0.0

RESULTANT JOINT DISPLACEMENTS - FREE JOINTS

JOINT	/------DISPLACEMENT------//			/------ROTATION------/		
	X DISP.	Y DISP.	Z DISP.	X ROT.	Y ROT.	Z ROT.
2			-0.0130727	0.0004468	0.0016696	
3			-0.0284589	0.0013657	0.0013657	
5			-0.0062760	0.0008877	0.0008877	
6			-0.0130727	0.0016696	0.0004468	

Example 9-4

Perform the analysis of the given plate in bending using the STRUDL program. The boundary condition is defined by the fixed support along one edge. The concentrated loads of 4^k, 8^k, 8^k, and 4^k are applied at the nodes 13, 14, 15, and 16 respectively. The thickness of the plate is 2 in. Young's modulus is 30,000 ksi and the Poisson ratio is 0.3. The geometry of the plate is given in Fig. 9-6.

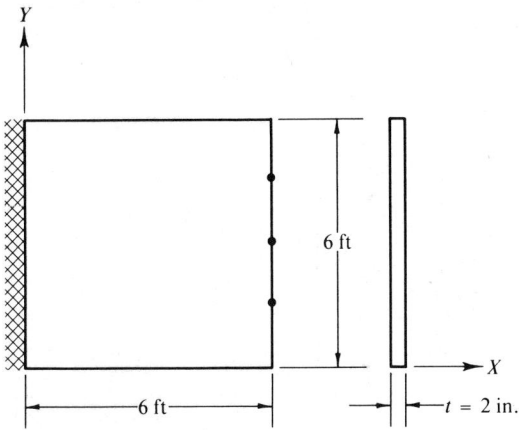

Figure 9-6

Solution

The mesh generation of the plate is done by 18 flat-plate triangular elements which are related to the analysis of bending problems. The elements and nodal points are shown in Fig. 9-7. The side which includes the nodes 1, 2, 3, and 4 is the fixed support. The concentrated loads of 4^k, 8^k, 8^k, and 4^k are applied at nodes 13, 14, 15, and 16, respectively. The solution of the finite element analysis of the problem is obtained by the STRUDL program. The input data is prepared by the use of proper commands.

```
STRUDL INPUT
STRUDL    'PROB 9-4'   'ANALYSIS OF A THIN PLATE IN
BENDING
$ USING TRIANGULAR ELEMENTS-TOTAL 18 ELEMENTS,
$    16 NODES
$ FORCES AT NODAL POINTS 13, 14, 15, 16 NEGATIVE
$    Z-DIRECTION
$ FIXED SUPPORT AT 1-2, 2-3, 3-4 EDGE OF PLATE.
$ ************************
TYPE BENDING
UNITS KIPS FEET
```

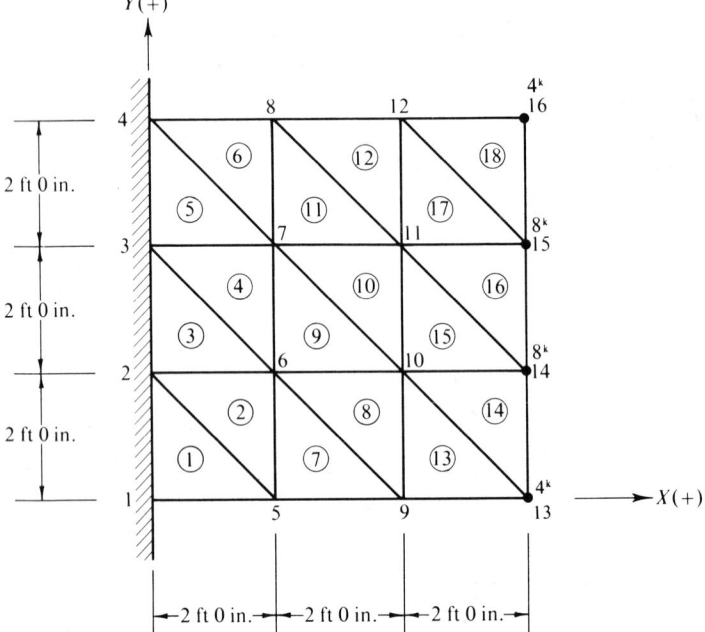

Figure 9-7

JOINT COORDINATES
```
1    0.   0.   S
2    0.   2.   S
3    0.   4.   S
4    0.   6.   S
5    2.   0.
6    2.   2.
7    2.   4.
8    2.   6.
9    4.   0.
10   4.   2.
11   4.   4.
12   4.   6.
13   6.   0.
14   6.   2.
15   6.   4.
16   6.   6.
```

ELEMENT INCIDENCES
```
1    1    5    2
2    2    5    6
3    2    6    3
4    3    6    7
```

TABLE 9-4 PROBLEM 9-4. ANALYSIS OF A THIN PLATE BENDING.

```
ACTIVE UNITS    INCH KIP  RAD DEGF SEC
ACTIVE STRUCTURE TYPE  PLATE  BENDING
ACTIVE COORDINATE AXES  X Y

LOADING - ONE           POINT

ELEMENT STRESSES
```

ELEMENT						
1	NODE 5	MXX	0.187410E 02	MYY	0.297443E 01	MXY 0.205752E 01
1	NODE 2	MXX	0.225237E 02	MYY	0.674713E 01	MXY -0.107460E-04
2	NODE 1	MXX	0.266640E 02	MYY	0.483764E 01	MXY 0.102876E 01
2	NODE 5	MXX	0.169722E 02	MYY	0.120570E 01	MXY 0.326523E 00
2	NODE 6	MXX	0.152904E 02	MYY	0.425689E 00	MXY 0.793130E 00
3	NODE 6	MXX	0.140891E 02	MYY	0.796672E 00	MXY 0.559825E 00
3	NODE 3	MXX	0.211840E 02	MYY	0.635523E 01	MXY 0.326550E 00
3	NODE 2	MXX	0.258717E 02	MYY	0.754923E 01	MXY -0.205849E-04
3	NODE 6	MXX	0.278304E 02	MYY	0.130017E 02	MXY 0.163265E 00
4	NODE 6	MXX	0.206396E 02	MYY	0.511028E 01	MXY 0.399503E 00
4	NODE 7	MXX	0.142086E 02	MYY	0.291173E 01	MXY -0.989816E-01
5	NODE 7	MXX	0.133105E 02	MYY	0.603297E 00	MXY -0.249245E 01
5	NODE 4	MXX	0.200059E 02	MYY	-0.602167E 00	MXY -0.399470E 00
5	NODE 3	MXX	0.205081E 02	MYY	0.804871E 00	MXY -0.151672E-04
6	NODE 4	MXX	0.260635E 02	MYY	0.724315E 01	MXY -0.212188E 01
6	NODE 7	MXX	0.206219E 02	MYY	0.661735E 01	MXY -0.688468E 00
7	NODE 8	MXX	0.145729E 02	MYY	0.389921E 01	MXY 0.140517E 01
7	NODE 9	MXX	0.641129E 01	MYY	0.343120E 01	MXY -0.225191E 01
7	NODE 6	MXX	0.185245E 02	MYY	0.169187E 01	MXY 0.793131E 00
8	NODE 5	MXX	0.168255E 02	MYY	0.611576E 00	MXY 0.385302E 00
8	NODE 6	MXX	0.164215E 02	MYY	0.842061E 00	MXY 0.213569E 00
8	NODE 10	MXX	0.815539E 01	MYY	0.912010E 01	MXY -0.184295E-01
8	NODE 10	MXX	0.691456E 01	MYY	-0.270829E 01	MXY -0.975680E-01
9	NODE 7	MXX	0.159554E 02	MYY	0.197256E 02	MXY 0.213569E 00
9	NODE 6	MXX	0.160658E 02	MYY	0.423500E 02	MXY 0.989820E-01
10	NODE 10	MXX	0.114742E 02	MYY	0.724984E 01	MXY -0.572900E-01
10	NODE 11	MXX	0.798645E 01	MYY	0.185136E 01	MXY -0.180957E 00
11	NODE 11	MXX	0.637578E 01	MYY	0.115119E 01	MXY -0.616916E 00
11	NODE 8	MXX	0.848922E 01	MYY	-0.445711E 01	MXY -0.121327E 00
12	NODE 7	MXX	0.165471E 02	MYY	0.236227E 01	MXY -0.180952E 00
12	NODE 8	MXX	0.175927E 02	MYY	0.505512E 01	MXY -0.688488E 00
12	NODE 8	MXX		MYY	0.674116E 01	MXY -0.434720E 00
12	NODE 8					MXY -0.253893E-01

231

TABLE 9-4 (continued).

```
  12  NODE 11   MXX -0.128762E 02   MYY  0.202470E 00   MXY -0.218614E 00
  12  NODE 12   MXX -0.122543E 00   MYY -0.106771E 00   MXY -0.122003E 00
  13  NODE 13   MXX -0.592488E 00   MYY -0.530316E-01   MXY -0.568053E 00
  13  NODE 10   MXX -0.475425E-01   MYY -0.435806E-01   MXY -0.187103E-01
  14  NODE  9   MXX -0.856012E-01   MYY -0.421103E 00   MXY -0.293384E 00
  14  NODE 10   MXX  0.949204E 00   MYY  0.478131E-01   MXY -0.199030E 00
  14  NODE 13   MXX  0.477118E-01   MYY  0.604587E-01   MXY -0.579735E 00
  14  NODE 14   MXX -0.235209E-01   MYY -0.181104E-01   MXY -0.389383E 00
  15  NODE 10   MXX -0.140426E 01   MYY -0.551460E-01   MXY -0.198989E 00
  15  NODE 11   MXX  0.409280E 01   MYY -0.943947E-01   MXY -0.616613E-01
  15  NODE 14   MXX  0.826066E 01   MYY -0.976178E 00   MXY -0.130326E 00
  16  NODE 11   MXX  0.875491E 01   MYY -0.456767E 01   MXY -0.270967E 00
  16  NODE 12   MXX  0.399997E 01   MYY -0.182267E 00   MXY -0.208373E 00
  16  NODE 15   MXX -0.246748E-01   MYY -0.166480E 01   MXY -0.239673E 00
  17  NODE 15   MXX -0.115230E 01   MYY -0.522408E 01   MXY -0.270778E 00
  17  NODE 12   MXX  0.333809E 01   MYY -0.123708E 01   MXY -0.218563E 00
  18  NODE 12   MXX -0.748246E 01   MYY -0.402560E 00   MXY -0.244671E 00
  18  NODE 15   MXX  0.102694E 02   MYY -0.569956E 01   MXY -0.276990E 00
  18  NODE 16   MXX -0.304957E 01   MYY -0.152227E 01   MXY -0.510078E 00
  18  NODE 18   MXX -0.117998E 01   MYY -0.107715E 01   MXY -0.116547E 00
```

ELEMENT STRAINS
ELEMENT

LOADING - ONE   POINT

RESULTANT JOINT DISPLACEMENTS - SUPPORTS

JOINT	/------DISPLACEMENT------/			/------ROTATION------/		
	X DISP.	Y DISP.	Z DISP.	X ROT.	Y ROT.	Z ROT.
1			0.0	0.0	0.0	
2			0.0	0.0	0.0	
3			0.0	0.0	0.0	
4			0.0	0.0	0.0	

RESULTANT JOINT DISPLACEMENTS - FREE JOINTS

JOINT	/------DISPLACEMENT------/			/------ROTATION------/		
	X DISP.	Y DISP.	Z DISP.	X ROT.	Y ROT.	Z ROT.
5			-0.2565037	-0.0032097	-0.0032097	0.0213861
6			-0.2942824	-0.0006232	-0.0226234	0.0226234
7			-0.2946186	-0.0005091	-0.0224691	0.0224691
8			-0.3609336	-0.0003107	-0.0213951	0.0213951
9			-0.3679306	-0.0031747	-0.0360437	0.0360437
10			-1.0134192	-0.0008426	-0.0361049	0.0361049
11			-1.0120913	-0.0009495	-0.0359188	0.0359188
12			-0.9619153	-0.0003495	-0.0355789	0.0355789
13			-1.9221382	-0.0002890	-0.0351722	0.0351722
14			-1.9564390	-0.0005329	-0.0419172	0.0419172
15			-1.9476728	-0.0013279	-0.0410131	0.0410131
16			-1.8913918	0.0029170	-0.0406882	0.0398928

5	3	7	4
6	4	7	8
7	5	9	6
8	6	9	10
9	6	10	7
10	7	10	11
11	7	11	8
12	8	11	12
13	9	13	10
14	10	13	14
15	10	14	11
16	11	14	15
17	11	15	12
18	12	15	16

```
UNITS INCHES
ELEMENT PROPERTIES
1 TO 18 TYPE 'CPT' THICKNESS 2.
CONSTANTS
E 30000. ALL
POISSON 0.30 ALL
LOADING 'ONE 'POINT'
JOINT LOADS
13 FORCE Z -4.0
14 FORCE Z -8.0
15 FORCE Z -8.0
16 FORCE Z -4.0
LOAD LIST ALL
STIFFNESS ANALYSIS
PRINT DATA
LIST STRESSES, STRAINS, DISPLACEMENTS, ALL
FINISH
```

STRUDL OUTPUT:
The results of the plate-bending analysis by the STRUDL program are shown in Table 9-4.

PROBLEMS

9-1 A simply supported beam shown in Fig. 9-8, supports a uniform load of 3.5 kips/ft. It spans 27 ft and its depth is 8 in. Young's modulus is 29,000 ksi and the Poisson ratio is 0.3. Perform the finite element analysis of the problem using the STRUDL program and constant-strain triangular elements.

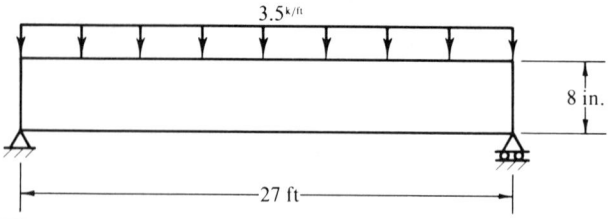

Figure 9-8

9-2 Perform the finite element analysis of the thin plate shown in Fig. 9-9 which acts under pure tension. Use constant-strain triangles to generate a rough mesh and then use STRUDL to obtain a solution. Young's modulus is 30,000 ksi and the Poisson ratio is 0.30. The plate is fixed along one of the short sides. The loading and geometry is shown in Fig. 9-9.

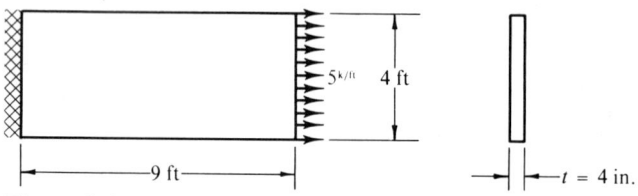

Figure 9-9

9-3 Perform the finite element analysis of the plate in bending using the STRUDL program. The plate is simply supported at two opposite edges. It is a square thin plate having a thickness of 5 in. Young's modulus is 30,000 ksi and the Poisson ratio is 0.3. It supports a uniform load of 2.5 kips/sq ft of surface area. Use flat-plate triangular elements to generate the mesh. The geometry and boundary conditions are shown in Fig. 9-10.

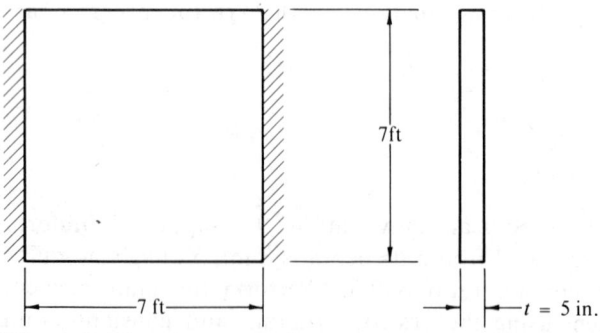

Figure 9-10

9-4 Repeat the analysis of the Problem 9-3 using a finer mesh and use STRUDL to obtain a new set of solutions. Then compare the results of the two solutions. Make an evaluation of your findings.

9-5 What are advantages and disadvantages of using ICES-STRUDL for the solution of analysis problems? Make your own engineering evaluation of the program.

REFERENCES

1. MANUAL NO. R67-56, Civil Engineering Department, Massachusetts Institute of Technology, Cambridge, Mass., September, 1967.
2. R. D. LOGCHER and G. M. STURMAN, "STRUDL-A Computer System for Structural Design," *Journal of Structures Division, American Society of Civil Engineers* **92,** ST6 (December 1966).
3. ICES-STRUDL II, Engineering User's Manual, **1, 2, 3,** Publication No. R68-91, R68-92, R68-93, Department of Civil Engineering, Massachusetts Institute of Technology, Cambridge, Mass., 1969.
4. STRESS: A Reference Manual, MIT Press, Cambridge, Mass., 1965.
5. J. CONNOR and G. WILL, "Computer-Aided Teaching of the Finite Element Displacement Method," School of Engineering, Massachusetts Institute of Technology, Cambridge, Mass., February 1969.

Applications of the Finite Element Method

INTRODUCTION

The value of the finite element method in civil engineering stems from its wide applications to the field of analysis. If the aid of a digital computer is available, almost any structural analysis problem can be formulated for entry into the computer to obtain a quick and accurate solution. In Chapter 4 a detailed discussion of the basic development of the finite-element theory was presented. In Chapter 5, longhand solutions of plane stress analysis problems were illustrated to secure a solid understanding of the method. Chapter 7 introduced the plate bending element and Chapter 9 illustrated the formulation and solution of plane stress and plate bending problems using a general purpose program, ICES-STRUDL.

The reader observes that user's point of view, the finite element method has two major sections:

a Idealization of the problem with logical mesh generation.
b Preparation of the input information to be used by a particular computer program to obtain the results of the analysis.

238 APPLICATION OF THE FINITE ELEMENT METHOD

To increase the ability of the reader to apply the finite-element method to practical problems, additional example problems will be discussed.

I would like to emphasize the need of the engineer to *thoroughly understand all relevant theory and behavior of structures before using any of the programs defined in this text.* The familiarity with the basic concepts of mathematics, computers, mechanics, materials and elasticity is necessary to avoid all potentially grave errors.

FORMULATION OF ANALYSIS PROBLEMS

The presentation of each problem has two parts: mesh generation and preparation of input information. The mesh generation scheme depends on the knowledge and experience of the engineer as well as on the nature of the problem to be analyzed. At the neighborhood of stress concentration areas a fine mesh is needed, and if highly accurate results are required the elements should be as small as practical.

Since the STRUDL program has been discussed and has also been used to obtain solutions of several problems, the input information for the problems of this section are prepared according to the commands of STRUDL. Another reason of doing so is that STRUDL is readily available at the majority of computer centers.

Examples 10-1 and 10-2 are plane stress analysis of a thin plate with a circular hole, using two different mesh generations. Geometry, boundary conditions, loading, and material properties are common to both problems. Constant-strain triangular elements are used for the idealization of the structure.

Examples 10-3, 10-4, and 10-5 are the analysis of a thin rectangular plate under the effect of tensile loads. The only difference between the three problems is the mesh generation. Since they are plane stress problems, constant strain triangular elements are used for their formulation.

Example 10-6 is the analysis of a rigid frame with nonprismatic members. To formulate for the plane stress analysis, constant strain triangular elements are used.

Example 10-7 is a simply supported beam which has two circular cutouts. It is analyzed using constant-strain triangular elements. Due to the symmetry, only one-half of the span is considered. A detailed mesh system is developed.

Example 10-8 is the analysis of a thin plate in bending. The formulation of the problem is done by using flat-plate triangular elements.

If the student has access to computer programs other than STRUDL, he is encouraged to prepare the input for these programs. This will be an excellent means to become familiar with these computer programming systems. Since the STRUDL input informations are already available, the student can prepare the related input cards for a quick computer solution for all of the eight analysis problems.

Each problem will be presented separately with generated mesh system and input information.

Example 10-1

Perform the analysis of the thin plate in plane stress. The plane has a circular hole of 12 in. in diameter at its center. The Young's modulus is 30,000 ksi and the Poisson ratio is 0.3. The plate is 4 ft square having a thickness of 1 in. The tensile load applied at opposite edges is 25 kips/linear ft. The geometry of the plate is shown in Fig. 10-1.

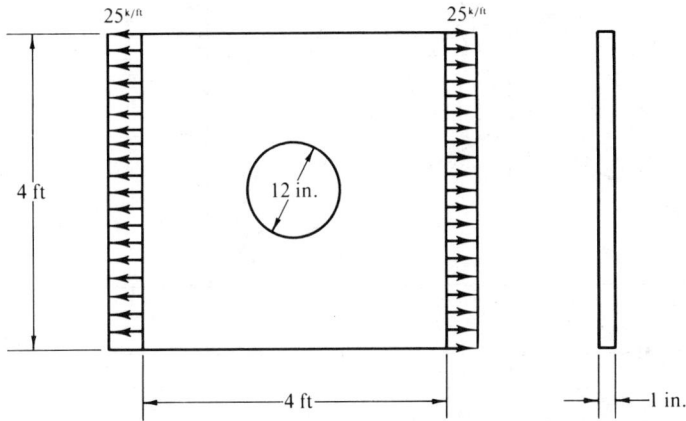

Figure 10-1

Solution
 a   Mesh generation.

This is a plane stress problem. Constant-strain triangular elements are used to generate the mesh for the analysis. Because of the symmetry only one-half of the plate is considered. The elements and nodes are shown in Fig. 10-2.

 b   STRUDL INPUT

The input information is prepared according to the STRUDL command discussed in Chapter 9. The student can use the given complete input data to obtain the results of the analysis. This will be an ideal exercise to fully comprehend the use of the finite-element method.

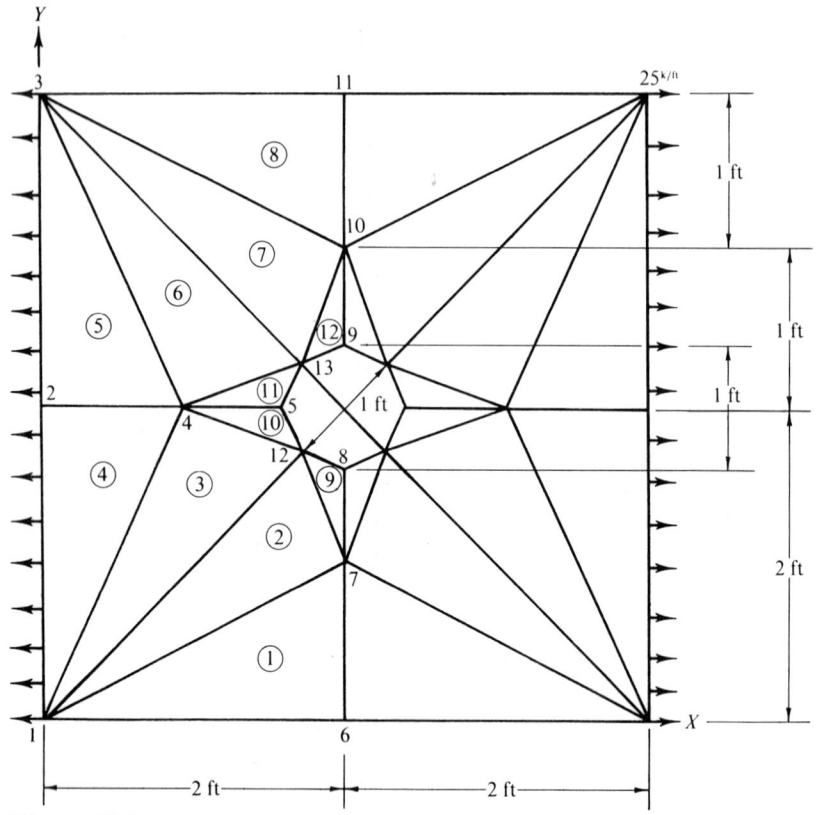

Figure 10-2

STRUDL 'EXA 10-1' 'PLATE IN PLANE STRESS WITH CIRCULAR HOLE'

TYPE PLANE STRESS
UNITS KIPS FEET
JOINT COORDINATES
1   0.   0.
2   0.   2.
3   0.   4.
4   1.   2.
5   1.5   2.
6   2.   0.   S
7   2.   1.   S
8   2.   1.5   S
9   2.   2.5   S
10 2.   3.   S
11 2.   4.   S
12 1.647   1.647
13 1.647   2.353

ELEMENT INCIDENCES
```
 1    1    6    7
 2    1    7   12
 3    1   12    4
 4    1    4    2
 5    2    4    3
 6    3    4   13
 7    3   13   10
 8    3   10   11
 9   12    7    8
10    4   12    5
11    4    5   13
12   13    9   10
```

JOINT RELEASES
6   7   8   9   10   11   FORCE   Y
UNITS KIPS INCHES
ELEMENT PROPERTIES
1 TO 12 TYPE 'CSTG' THICKNESS 1.
CONSTANTS
E 30000. ALL
POISSON 0.3 ALL
LOADING 'ENDS' 'TO SIMULATE LINEAR END FORCES'
JOINT LOADS
1   2   3   FORCE X -33.33
PRINT DATA
STIFFNESS ANALYSIS
LIST DISPLACEMENTS, STRESSES, STRAINS, ALL
FINISH

Example 10-2

Reanalyze the thin plate described in Problem 10-1 by the finite-element method, using twice as many elements used previously. All the other data will be the ones given for Problem 10-1. The geometry and loading are described in Fig. 10-1.

Solution

a Mesh generation

Constant-strain triangular elements are used to generate the mesh for the problem. The elements and nodes are given in Fig. 10-3.

b STRUDL INPUT

STRUDL   'EXA 10-2'   'EXA 10-1 REPEATED WITH TWICE AS MANY CSTG ELEMENTS'

242 APPLICATION OF THE FINITE ELEMENT METHOD

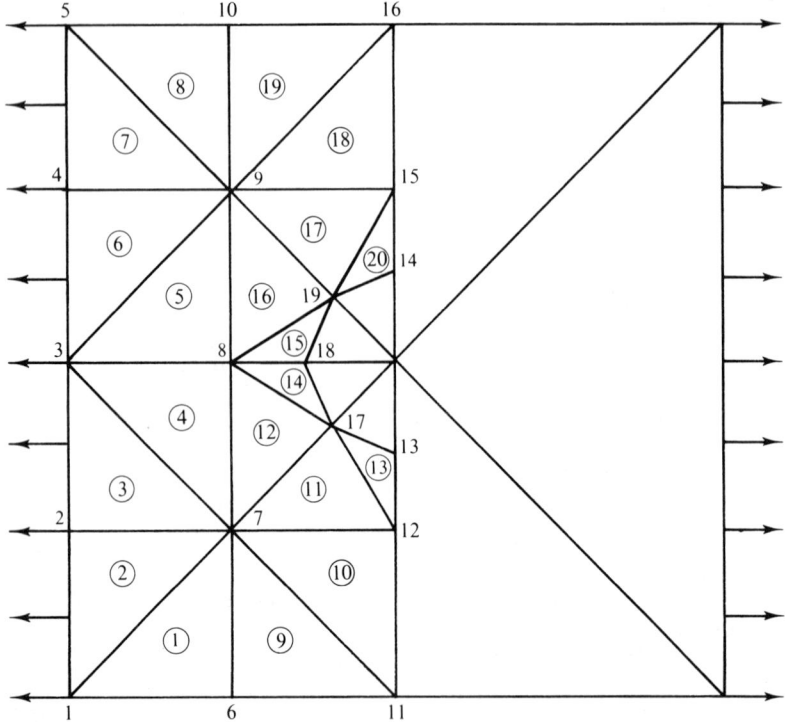

Figure 10-3

TYPE PLANE STRESS
UNITS KIPS FEET
JOINT COORDINATES

1	0.	0.	
2	0.	1.	
3	0.	2.	
4	0.	3.	
5	0.	4.	
6	1.	0.	
7	1.	1.	
8	1.	2.	
9	1.	3.	
10	1.	4.	
11	2.	0.	S
12	2.	1.	S
13	2.	1.5	S
14	2.	2.5	S
15	2.	3.	S
16	2.	4.	S
17	1.647	1.647	
18	1.5	2.	
19	1.647	2.353	

ELEMENT INCIDENCES
```
 1    1    6    7
 2    1    7    2
 3    2    7    3
 4    3    7    8
 5    3    8    9
 6    3    9    4
 7    4    9    5
 8    5    9   10
 9    7    6   11
10    7   11   12
11    7   12   17
12    7   17    8
13   12   13   17
14    8   17   18
15    8   18   19
16    8   19    9
17    9   19   15
18    9   15   16
19    9   16   10
20   19   14   15
```

JOINT RELEASES
11 TO 16 FORCE Y
UNITS KIPS INCHES
ELEMENT PROPERTIES
1 TO 20 TYPE 'CSTG' THICKNESS 1.
CONSTANTS
E 30,000 ALL
POISSON 0.3 ALL
LOADING 'ENDS' 'TO SIMULATE LINEAR END FORCES'
JOINT LOADS
1 TO 5 FORCE X -20.
PRINT DATA
STIFFNESS ANALYSIS
LIST DISPLACEMENT STRESSES STRAINS ALL
FINISH

Comparative Study

   a Time for data preparation
        Coarse mesh (Problem 10-1)   20 min
        Fine mesh (Problem 10-2)   25 min
   b Computer run-time
        Coarse mesh   54.43 sec
        Fine mesh   56.83 sec

Example 10-3

Perform the analysis of a thin rectangular plate of 20 ft long, length, 10 ft wide, and 1 in. thick by the finite-element method. It is loaded by

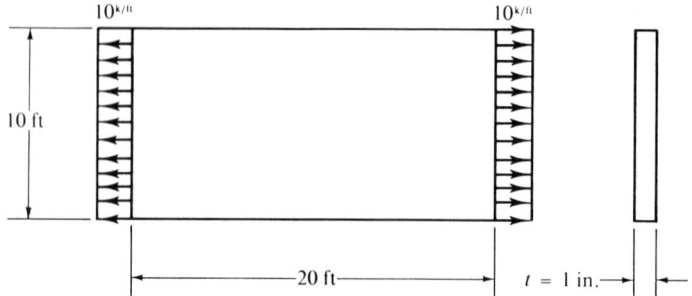

Figure 10-4

uniform tensile loads along the 10-ft edges. The Young's modulus is 30,000 ksi and the Poisson ratio is 0.3. The geometry and loading are shown in Fig. 10-4.

Solution

a Mesh generation

This is a plane stress analysis problem, therefore constant-strain triangular elements are used for the mesh generation of the problem. Because of the symmetry, only one-half of the structure is considered. The topology of the plate is shown in Fig. 10-5. This mesh has only eight elements.

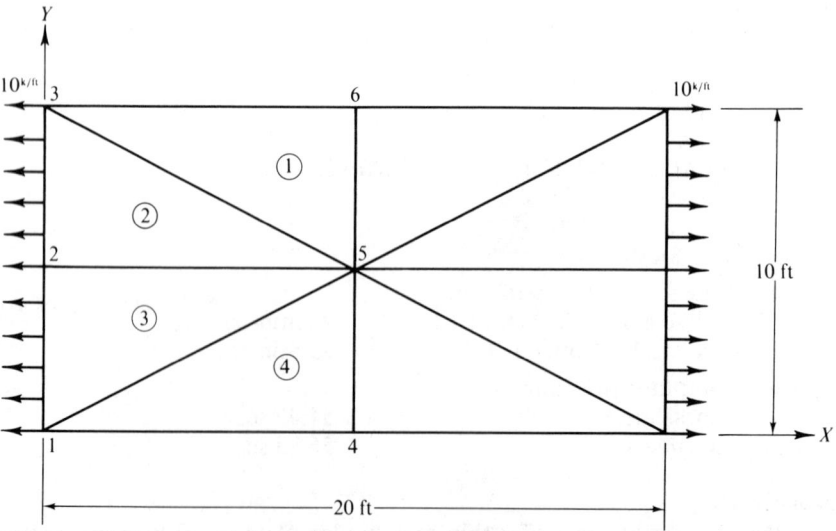

Figure 10-5

b STRUDL INPUT

STRUDL   'EXA 10-3'   'PLANE STRESS ANALYSIS OF A PLATE WITH 8 ELEMENTS'

TYPE PLANE STRESS
UNITS KIPS FEET
JOINT COORDINATES
1   0.   0.
2   0.   5.
3   0.   10.
4   10.   0.   S
5   10.   5.   S
6   10.   10.   S

ELEMENT INCIDENCES
1   3   5   6
2   2   5   3
3   1   5   2
4   1   4   5
$ MAKE USE OF SYMMETRY TO ANALYSIS ONLY 1/2 OF PLATE
$ RECOGNIZE SYMMETRY IN SPECIFIED RELEASES
JOINT RELEASES
4   6   FORCE Y
UNITS KIPS INCHES
ELEMENT PROPERTIES
1 TO 4 TYPE 'CSTG' THICKNESS 1.
CONSTANTS
E 30000. ALL
POISSON 0.3 ALL
LOADING 'END' 'TO SIMULATE-LINEAR LOADING ON ONE END'
JOINT LOADS
1   2   3   FORCE X -33.33
LOADING LIST ALL
PRINT DATA
STIFFNESS ANALYSIS
LIST DISPLACEMENTS, STRESSES, STRAINS, ALL
FINISH

Example 10-4

Reanalyze Problem 10-3 using twice as many triangular elements. Use all the other necessary data as in Problem 10-3.

Solution

a Mesh generation

The mesh has 16 constant-strain triangular elements which is twice as many as there were in Problem 10-3. This condition will improve the

246 APPLICATION OF THE FINITE ELEMENT METHOD

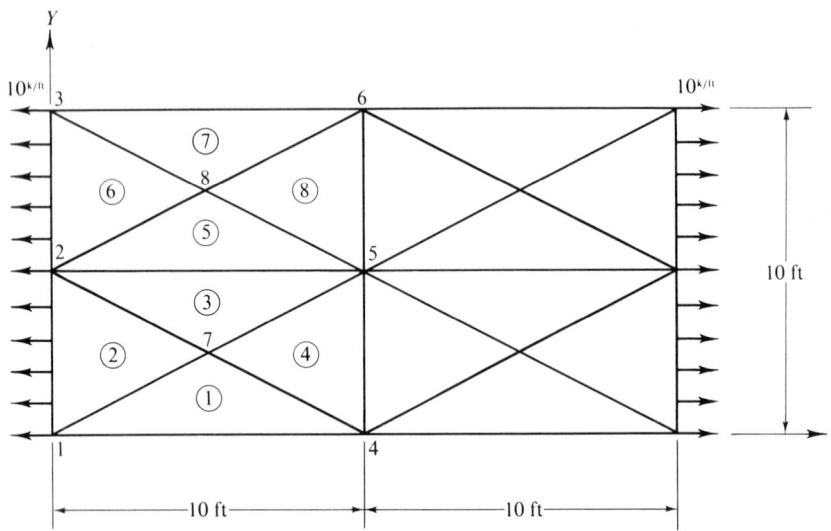

Figure 10-6

accuracy of the results. The new idealization of the structure is shown in Fig. 10-6.

b STRUDL INPUT

STRUDL   'EXA 10-4'   'PLANE STRESS ANALYSIS OF A PLATE WITH 16 ELEMENTS'

TYPE PLANE STRESS
UNITS KIPS FEET
JOINT COORDINATES
1   0.   0.
2   0.   5.
3   0.   10.
4   10.   0.   S
5   10.   5.   S
6   10.   10.   S
7   5.   2.5
8   5.   7.5

ELEMENT INCIDENCES
1   1   4   7
2   1   7   2
3   2   7   5
4   4   5   7
5   2   5   8
6   2   8   3

FORMULATION OF ANALYSIS PROBLEMS 247

```
7    3    8    6
8    5    6    8
$ MAKE USE OF SYMMETRY TO ANALYSIS ONLY 1/2 OF THE PLATE
$ RECOGNIZE SYMMETRY IN SPECIFIED RELEASES
JOINT RELEASES
4    6    FORCE Y
UNITS KIPS INCHES
ELEMENT PROPERTIES
1 TO 8 TYPE 'CSTG' THICKNESS 1.
CONSTANTS
E 30,000. ALL
POISSON 0.3 ALL
LOADING 'END''TO SIMULATE END LOADING'
JOINT LOADS
1    2    3    FORCE X -33.33
LOADING LIST ALL
PRINT DATA
STIFFNESS ANALYSIS
LIST DISPLACEMENTS, STRESSES, STRAINS, ALL
FINISH
```

Example 10-5

Reanalyze Problem 10-3 using 32 constant-strain triangular elements. Use all other necessary data as given in Problem 10-3. Because of the finer mesh used, the results will tend to converge to the exact solution of the analysis. This will illustrate the concept of convergence discussed in Chapter 3.

Solution

    a Mesh generation

The mesh generation has 32 triangular elements as shown in Fig. 10-7.

    b STRUDL INPUT

```
STRUDL    'EXA 10-5'    'PLANE STRESS ANALYSIS OF A PLATE
                        WITH 32 ELEMENTS'

TYPE PLANE STRESS
UNITS KIPS FEET
JOINT COORDINATES
   1    0.      0.
   2    0.      2.5
   3    0.      5.
```

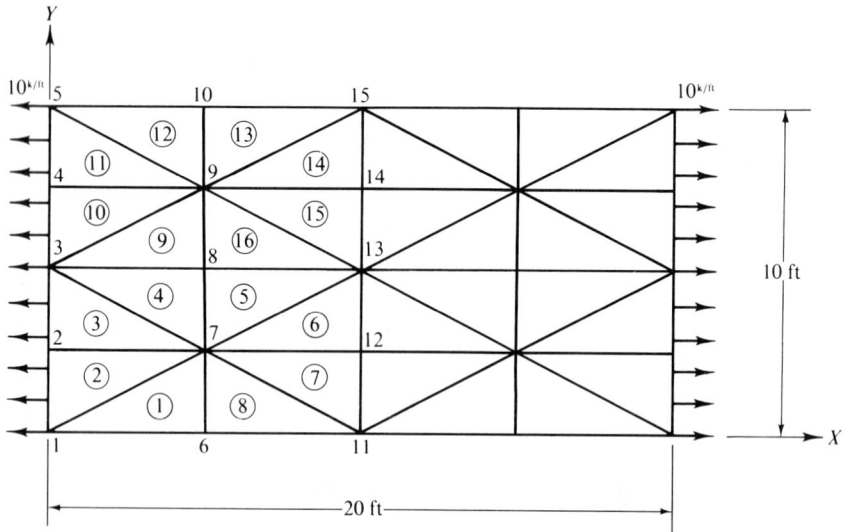

Figure 10-7

4	0.	7.5	
5	0.	10.	
6	5.	0.	
7	5.	2.5	
8	5.	5.	
9	5.	7.5	
10	5.	10.	
11	10.	0.	S
12	10.	2.5	S
13	10.	5.	S
14	10.	7.5	S
15	10.	10.	S

ELEMENT INCIDENCES

1	1	6	7
2	1	7	2
3	2	7	3
4	3	7	8
5	8	7	13
6	7	12	13
7	7	11	12
8	7	6	11
9	3	8	9
10	3	9	4
11	4	9	5
12	5	9	10
13	10	9	15
14	9	14	15

```
15    9    13   14
16    9    8    13
$ MAKE USE OF SYMMETRY TO ANALYSIS ONLY 1/2 OF THE PLANE
$ RECOGNIZE SYMMETRY IN SPECIFIED RELEASES
JOINT RELEASES
11   12   14   15 FORCE Y
UNITS KIPS INCHES
ELEMENT PROPERTIES
1 TO 16 TYPE 'CSTG' THICKNESS 1.
CONSTANTS
E 30,000. ALL
POISSON 0.3 ALL
LOADING 'END' 'TO SIMULATE-LINEAR LOADING ON ONE END'
JOINT LOADS
1 TO 5 FORCE X -20.
LOADING LIST ALL
PRINT DATA
STIFFNESS ANALYSIS
LIST DISPLACEMENTS STRESSES STRAINS ALL
FINISH
```

Comparative Study

a Time for data preparation
   8 elements   10 min
   16 elements   11 min
   32 elements   16 min

b Computer run-time
   8 elements   40.71 sec
   16 elements   46.14 sec
   32 elements   56.41 sec

Example 10-6

Perform the stiffness finite element analysis of the rigid frame with nonprismatic members. The loading, caused by wind, has an intensity 2 kips/linear ft along one elevation of the frame. The modulus of elasticity is 30,000 ksi and the Poisson ratio is 0.3. The thickness of the plate of the frame members is 1 in. The geometry of the frame is shown in Fig. 10-8.

Solution

a Mesh generation

This problem is a plane stress analysis, therefore constant-strain triangular elements are used to develop the mesh. The elements and nodes are shown in Fig. 10-9.

250 APPLICATION OF THE FINITE ELEMENT METHOD

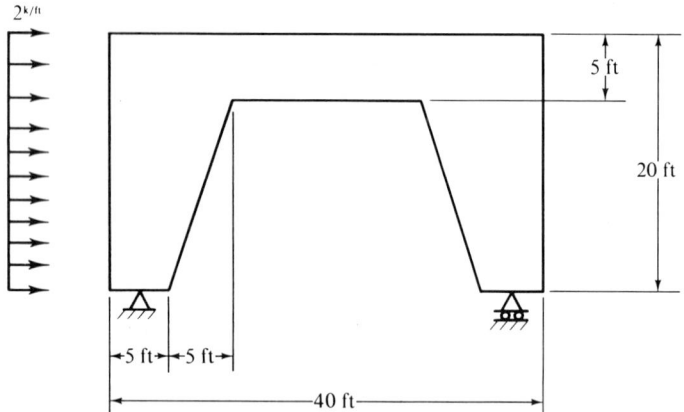

Figure 10-8

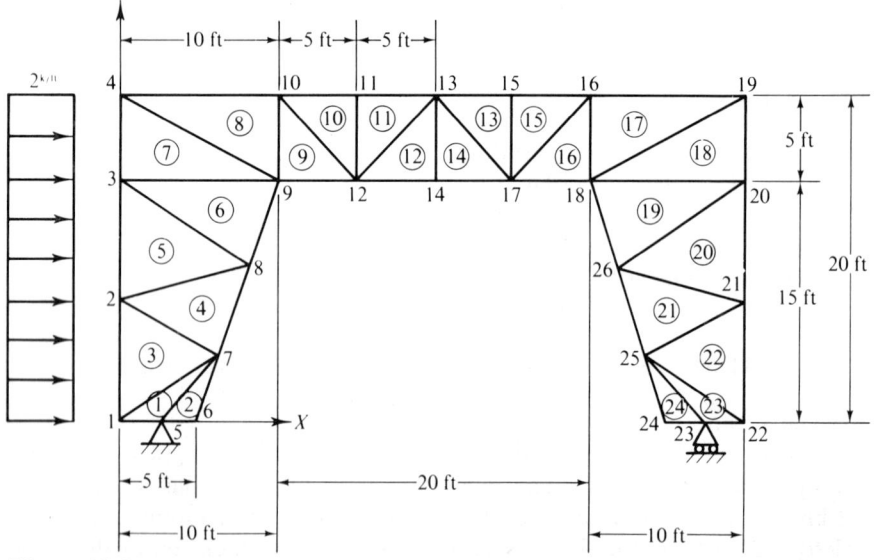

Figure 10-9

b STRUDL INPUT

STRUDL   'EXA 10-6'   'ANALYSIS OF A RIGID FRAME'

TYPE PLANE STRESS
UNITS FEET KIPS
JOINT COORDINATES
  1   0.   0.
  2   0.   7.5

3	0.	15.	
4	0.	20.	
5	2.5	0.	S
6	5.	0.	
7	6.667	5.	
8	8.333	10.	
9	10.	15.	
10	10.	20.	
11	15.	20.	
12	15.	15.	
13	20.	20.	
14	20.	15.	
15	25.	20.	
16	30.	20.	
17	25.	15.	
18	30.	15.	
19	40.	20.	
20	40.	15.	
21	40.	7.5	
22	40.	0.	
23	37.5	0.	
24	3.5	0.	
25	33.333	5.	
26	31.667	10.	

ELEMENT INCIDENCES

1	1	5	7
2	5	6	7
3	1	7	2
4	2	7	8
5	2	8	3
6	3	8	9
7	3	9	4
8	4	9	10
9	9	12	10
10	10	12	11
11	11	12	13
12	12	14	13
13	13	17	15
14	13	14	17
15	15	17	16
16	16	17	18
17	16	18	19
18	18	20	19
19	18	26	20
20	20	26	21
21	26	25	21
22	25	22	21
23	25	23	22
24	24	23	25

JOINT RELEASES
23 FORCE X
UNITS KIPS INCHES
ELEMENT PROPERTIES
1 TO 24 TYPE 'CSTG' THICKNESS 1.
CONSTANTS
E 30000. ALL
POISSON 0.3 ALL
UNITS KIPS FEET
LOADING 1 'SIMULATE LOAD ON LEFT'
JOINT LOADS
1   2   3   4   FORCE X 10.
PRINT DATA
STIFFNESS ANALYSIS
LIST STRESSES, DISPLACEMENT, ALL
FINISH

Example 10-7

Perform the finite element analysis of the thin deep beam shown in Fig. 10-10. It spans 15 ft and it is 5 ft deep. There are two circular cutouts, each having a diameter of 20 in. The modulus of elasticity of the

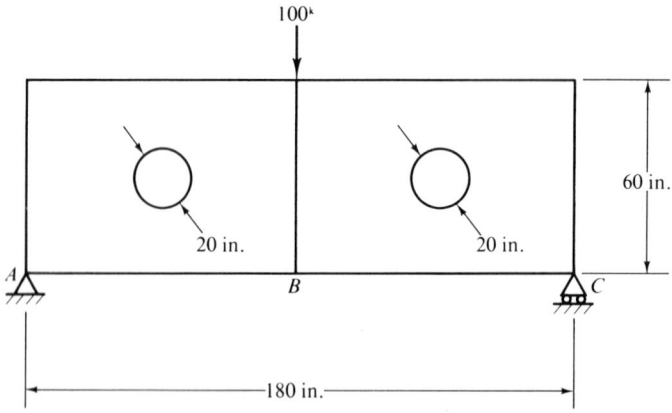

Figure 10-10

material is 30,000 ksi and the Poisson ratio is 0.3. Assume a thickness for the beam of 1 in. The geometry and loading of the beam is shown in Fig. 10-10.

Solution

a Mesh generation

Because of the symmetry of the structure, only one-half of the beam is considered. This is a plane stress problem that will use constant-strain

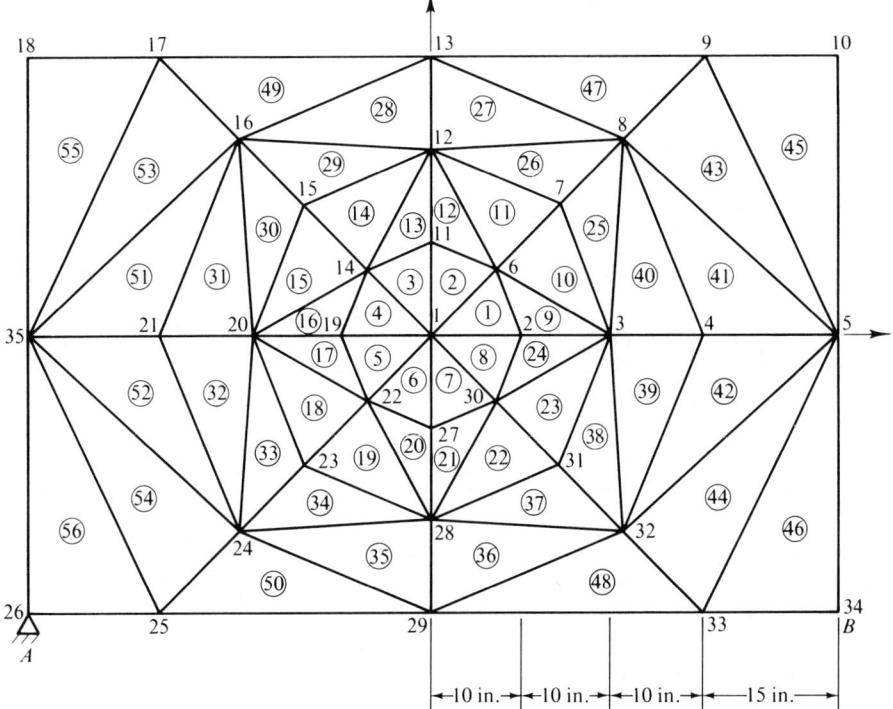

Figure 10-11

triangular elements. A comparatively fine mesh generation is shown in Fig. 10-11, which describes all the elements and nodes.

b STRUDL INPUT

```
STRUDL    'EXA 10-7'    'ANALYSIS OF A THIN BEAM'
TYPE PLANE STRESS
$ DUE TO SYMMETRY ONLY ½ OF BEAM IS USED

JOINT COORDINATES
 1      0.       0.     S
 2     10.       0.     S
 3     20.       0.
 4     30.       0.
 5     45.       0.     S
 6      7.06     7.06   S
 7     14.12    14.12
 8     21.18    21.18
 9     28.24    28.24
10     45.      30.     S
```

12	0.	20.	
13	0.	30.	
14	−7.06	7.06	S
15	−14.12	14.12	
16	−21.18	21.18	
17	−28.24	28.24	
18	−45.	30.	
19	−10.	0.	
20	−20.	0.	
21	−30.	0.	
35	−45.	0.	
26	−45.	−30.	S
22	−7.06	−7.06	S
23	−14.12	−14.12	
24	−21.18	−21.18	
25	−28.24	−28.24	
27	0.	−10.	
28	0.	−20.	
29	0.	−30.	
30	7.06	−7.06	S
31	14.12	−14.12	
32	21.18	−21.18	
33	28.24	−28.24	
34	45.	−30.	S

ELEMENT INCIDENCES

1	1	2	6
2	1	6	11
3	1	11	14
4	1	14	19
5	1	19	22
6	1	22	27
7	1	27	30
8	1	30	2
9	2	3	6
10	6	3	7
11	6	7	12
12	6	12	11
13	11	12	14
14	14	12	15
15	14	15	20
16	19	14	20
17	19	20	22
18	22	20	23
19	22	23	28
20	22	28	27
.	.	.	.
.	.	.	.
56	17	18	35

```
UNITS KIPS INCHES
JOINT RELEASES
1    2    6    11    14    19    22    27    30    FORCE X Y
34   5    10        FORCE Y
26                  FORCE X

ELEMENT PROPERTIES
1 TO 56 TYPE 'CSTG' THICKNESS 1.
CONSTANTS
E 30000. ALL
POISSON 0.3 ALL
LOADING 1 'POINT LOAD'
JOINT LOADS
10 FORCE Y -50
PRINT DATA
STIFFNESS ANALYSIS
LIST DISPLACEMENTS, STRESSES, STRAINS, ALL
FINISH
```

Example 10-8

Perform the finite element analysis of a plate in bending whose geometric dimensions, boundary conditions and loading are shown in Fig. 10-12. The plate is 3 in. thick and carries a uniform load of 200 lb/square feet area. Young's modulus is 29,000 ksi and the Poisson ratio is 0.3.

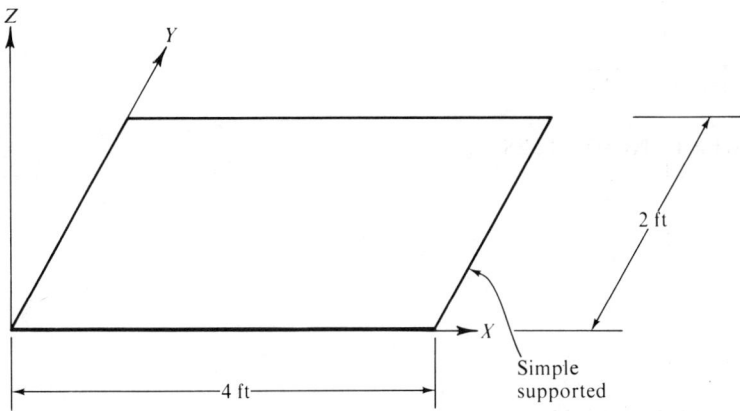

Figure 10-12

Solution

a Mesh generation

The mesh generation uses eight flat-plate triangular elements as shown in Fig. 10-13.

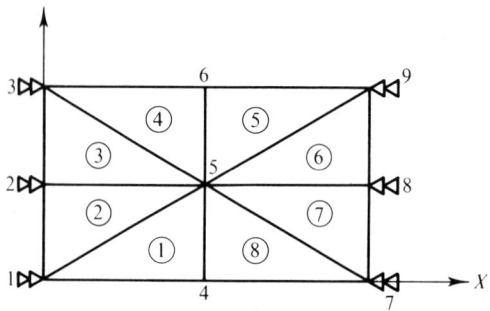

Figure 10-13

b STRUDL INPUT

STRUDL   'EXA 10-8'   'ANALYSIS OF A PLATE IN BENDING'

TYPE BENDING
UNITS KIPS FEET
JOINT COORDINATES
1   0.   0.   S
2   0.   1.   S
3   0.   2.   S
4   2.   0.
5   2.   1.
6   2.   2.
7   4.   0.   S
8   4.   1.   S
9   4.   2.   S

ELEMENT INCIDENCES
1   1   4   5
2   2   1   5
3   3   2   5
4   3   5   6
5   6   5   9
6   9   5   8
7   5   7   8
8   4   7   5

UNITS KIPS INCHES
ELEMENT PROPERTIES
1 TO 8 TYPE 'CPT' THICKNESS 3.
CONSTANTS
E 29000. ALL
POISSON 0.3 ALL
JOINT RELEASES
1   2   3   7   8   9   MOMENT X Y
UNITS KIPS FEET

```
LOADING 'ONE' 'UNIFORM'
ELEMENT LOADS
1 TO 8 SURFACE FORCES GLOBAL PZ -.2
UNITS INCHES
PRINT DATA
STIFFNESS ANALYSIS
LIST DISPLACEMENTS, STRESSES, ALL
FINISH
```

As mentioned before, the use of the prepared input to obtain computer solutions to the eight problems is left to the student as an exercise.

ADDITIONAL APPLICATIONS OF THE FINITE ELEMENT METHOD

The possible applications of the finite element method go far beyond the structural analysis problems that have been discussed until present. The presentations that follow are illustrative in nature, having the simple aim of introducing new horizons for imaginative engineers. Since their detailed formulation and solution are not within the scope of this book, only a preliminary mesh generation for each problem will be given. Students of civil engineering are encouraged to develop a logical and more detailed idealization of the problems of their interest and then prepare the necessary input data to be entered in a computer. Therefore our discussion in this section will be descriptive in nature rather than an exact formulation of the problem.

The four typical problems considered in this presentation are

  a Analysis of a gravity dam
  b Slope-stability analysis
  c Tunnel analysis
  d Flow through channels

Now let us discuss each one of these problems individually and generate a mesh for them.

A. Analysis of a Gravity Dam

The geometry and possible loading condition of a gravity dam are shown in Fig. 10-14. The following information is related to the description of the problem.

    Area (ABC) = 200 ft^2
    Area (dam) = 1200 ft^2

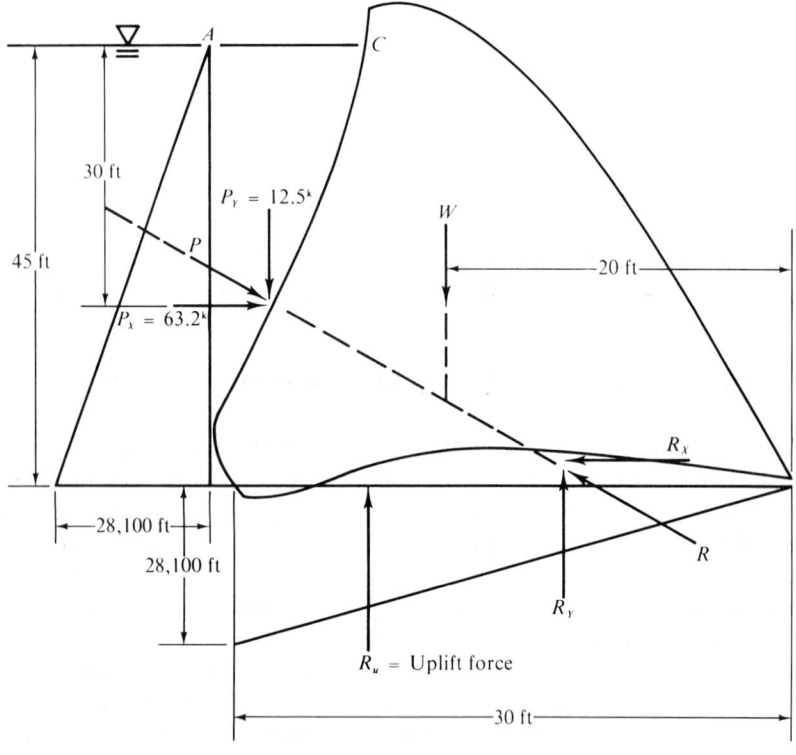

Figure 10-14

Weight of concrete = 150 lb/ft³
Height of dam = 60 ft

The loads on the dam can be summarized in five items.

1. $P_x = X$ = component of the hydrostatic pressure distributed at each nodal point (from 1 to 8).
2. $P_y = Y$ = component of the hydrostatic pressure due to the water in ABC distributed at each nodal point (from 1 to 8).
3. $R_x = X$ = component of the reaction between the foundation and the base of the dam structure distributed at nodal points (nodes 1, 11 to 22).
4. $R_y = Y$ = component of the reaction between the foundation and the base of the dam structure distributed at nodal points (nodes 1, 11 to 22).

5 The dead load of the dam distributed to all internal nodes. The analysis of the dam can be performed as a plane stress problem which consists of a cross section of the structure. Then constant-strain triangular elements are used for the formulation of the analysis. STRUDL problem will be applicable to obtain the solution. The engineer who is preparing the input data should define the boundary conditions as compatibile with the true conditions. The dead load of the dam needs to be distributed to all internal nodal points. Externally applied loads should be transmitted to the external nodes. An illustrative mesh generation is shown in Fig. 10-15.

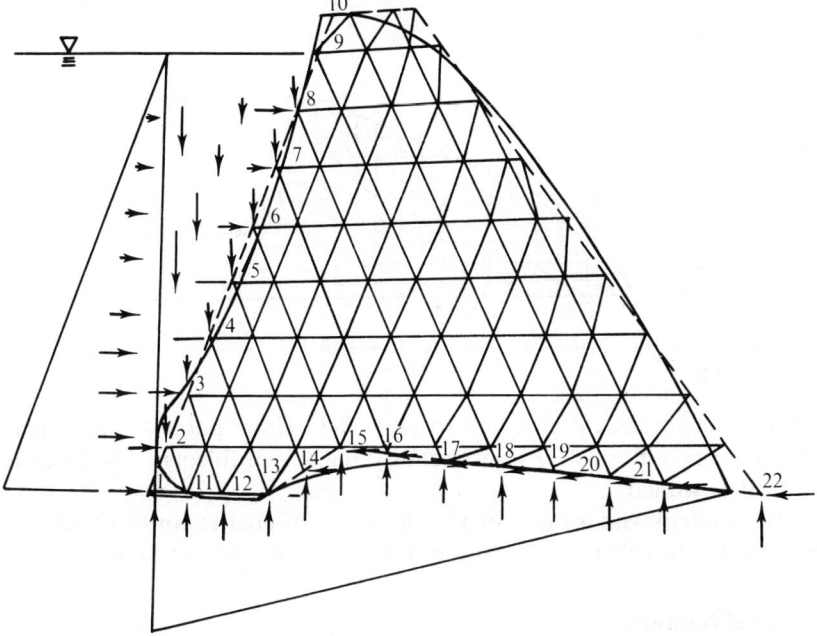

Figure 10-15

B. Slope-Stability Analysis

Slope-stability analysis is an important problem in the study of soil mechanics and foundation engineering. The finite element method can be a great asset in their analysis. The advantage of the method stems from its ability to describe the material property variations, since each element can present a different material. A computer program which can define the behavior of soil mass can effectively use this method to reach reasonable results. It is believed that with the aid of finite element method and relevant computer programs, soil mechanics and foundation engineering can move one step further in securing more theoretical flair to their

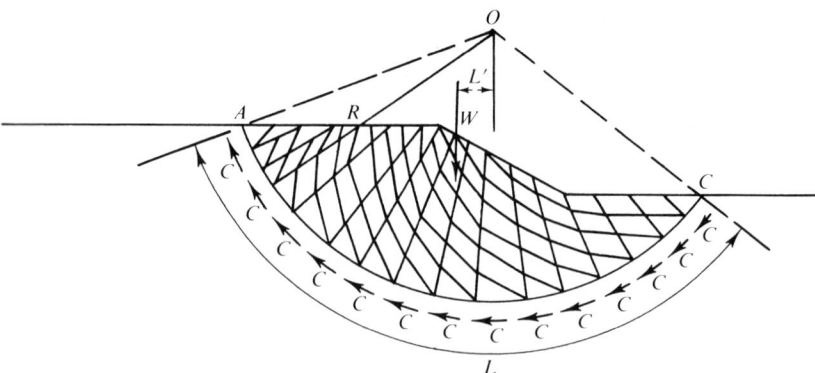

Figure 10-16

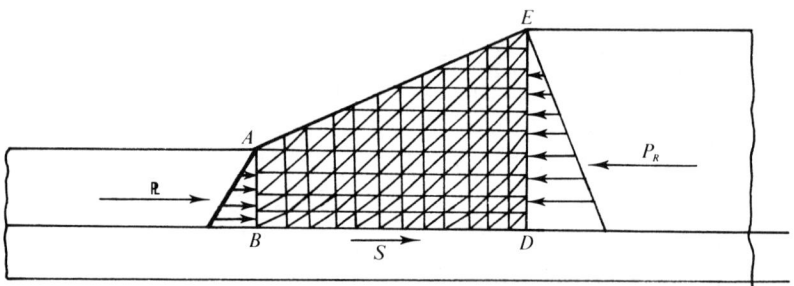

Figure 10-17

analysis approaches. Figures 10-16 and 10-17 show possible mesh generations for two illustrative slope-stability analysis problems. The force E is to be determined.

The student who is interested in this topic of civil engineering should explore other possible applications of finite element method to this area.

C. Tunnel Analysis

Tunnel analysis is a major problem in the field of civil engineering. One encounters it in transportation network engineering, construction engineering, water resources engineering, and mining engineering. Each tunnel analysis problem will have its particular complications due to the soil properties of the surrounding media, the loading conditions, climactic effects, and the method of performing the actual drilling and construction. All these factors will influence the stress flow and concentrations around the boundaries and neighborhood of the tunnel itself. These stress values have to be determined to critical accuracy to avoid any possible failure. For the theoretical study of this problem, the finite element technique is very useful. The solution of a stress problem related to a continuum presents some difficulties if classical methods are used. The finite element technique will facilitate the task to great extent. A mesh style will be developed depending on the accuracy desired for its solution. More elements should be placed around the tunnel, since they are the most critical stress areas. The mesh field is numbered properly. Then, according to the computer program to be used, the INPUT data is prepared.

Figure 10-18 illustrates a tunnel analysis problem. A random mesh generation is developed for the half of soil media due to the assumed sym-

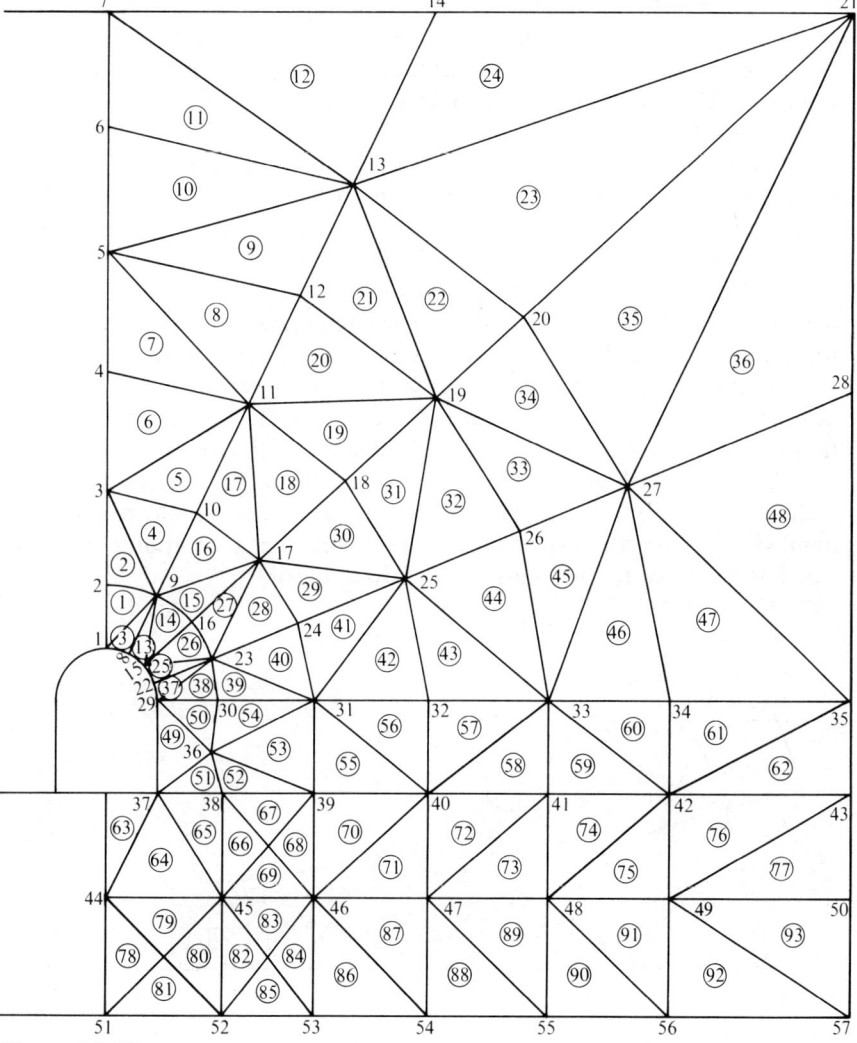

Figure 10-18

262 APPLICATION OF THE FINITE ELEMENT METHOD

metry of the loading, material uniformity and the geometry of the structure. An engineer who is familiar with tunnel analysis will be able to formulate the problem efficiently. The applications of the finite element technique to mining engineering and the field of rock mechanics are of major importance. Once the few constants related to the properties of the surrounding materials have been determined, this method will yield quick, painless, and accurate results.

D. Flow Through a Channel

Another area of the application of the finite element method is the study of problems related to hydraulics. The flow of water through a channel with introduced restrictions will present a difficult problem of analysis in determining the velocity of each water particle. The finite-element mesh generation will divide the cross section of the flow into very small parts whose behavior will define the behavior of the complete flow. The governing equations which will use finite element technique assumptions will be those relating to the theory of hydraulics. The boundary conditions of the flow problem need to be correctly defined in order to reach a reasonably correct solution. The engineer who is familiar with this topic can properly describe the prevailing boundaries. Since water is an incompressible medium, the elements defined should have special characteristics. Even though there are difficulties in the formulation, a solution can be obtained with further detailed study of the topic.

A typical mesh generation of a flow problem, illustrative in nature, is shown in Fig. 10-19. The elements and the nodes should be properly numbered. Then the INPUT data is prepared for a computer program which is pertinent to the solution of a flow analysis.

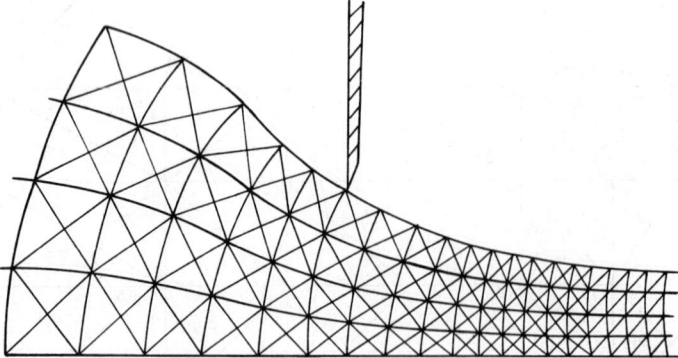

Figure 10-19

Supplementary Readings

I. LIST OF BOOKS

J. E. AKIN, D. L. FENTON, and W. C. T. STODDART, *The Finite Element Method—A Bibliography of Its Theory and Applications,* Report EM72-1, University of Tennessee, Knoxville, Tenn., February 1972.

R. H. GALLAGHER, *A Correlation Study of Methods of Matrix Structural Analysis.* Agardograph 69, 1962.

J. M. GERE and W. WEAVER, JR., *Matrix Algebra for Engineers.* Princeton, N. J.: Van Nostrand, 1967.

H. C. MARTIN, *Introduction to Matrix Methods of Structural Analysis.* New York: McGraw-Hill, 1966.

J. L. MEEK, *Matrix Structural Analysis.* New York: McGraw-Hill, 1971.

J. T. ODEN, *Finite Elements of Nonlinear Continua.* New York: McGraw-Hill, 1972.

J. S. PREZEMIENIECKI, *Theory of Matrix Structural Analysis.* New York: McGraw-Hill, 1968.

M. F. RUBINSTEIN, *Matrix Computer Analysis of Structures.* Englewood Cliffs, N. J.: Prentice-Hall, 1966.

M. F. RUBINSTEIN, *Structural Systems-Statics, Dynamics and Stability.* Englewood Cliffs, N. J.: Prentice-Hall, 1970.

M. G. SALVADORI and M. L. BARON, *Numerical Methods in Engineering.* Englewood Cliffs, N. J.: Prentice-Hall, 1961.

I. S. SOKOLNIKOFF, *Mathematical Theory of Elasticity.* New York: McGraw-Hill, 1956.

S. TIMOSHENKO and J. N. GOODIER, *Theory of Elasticity.* New York: McGraw-Hill, 1970.

OKTAY URAL, *Matrix Methods and Use of Computers in Structural Engineering.* Scranton, Pa.: International Textbook, 1971.

C. R. WYLIE, *Advanced Engineering Mathematics.* New York: McGraw-Hill, 1960.

O. C. ZIENKIEWICZ and Y. K. CHEUNG, *The Finite Element Method in Structural and Continuum Mechanics.* New York and London: McGraw-Hill, 1967.

O. C. ZIENKIEWICZ and G. G. HOLLISTER, *Stress Analysis.* New York: Wiley, 1965. "The Finite Element Method in Structural Mechanics," Chapter 7, by R. W. Clough; "Finite Element Procedures in the Solution of Plate and Shell Problems," Chapter 8, by O. C. Zienkiewicz, and Y. K. Cheung; "Displacement and Equilibrium Models in the Finite Element Method," Chapter 9, by B. Fraeijs de Venbeke.

II. LIST OF PAPERS

A. ADINI, "Analysis of Shell Structure by the Finite Element Method," Ph.D. Dissertation, Department of Civil Engineering, University of California, Berkeley, Calif., 1961.

A. ADINI and R. W. CLOUGH, "Analysis of Plate Bending by the Finite Element Method," Report submitted to the National Science Foundation, Great G7397, March 1960.

J. H. ARGYRIS, "Matrix Displacement Analysis of Anisotropic Shells by Triangular Elements," *Journal of the Royal Aeronautical Society* **69** (November 1965).

J. H. ARGYRIS, "Triangular Elements with Linearly Varying Strain for the Matrix Displacement Method," *Journal of the Royal Aeronautical Society, Technical Note* (October 1965).

G. P. BAZELEY, Y. K. CHEUNG, B. M. IRONS, and O. C. ZIENKIEWICZ, "Triangular Elements in Plate Bending—Conforming and Non-Conforming Solutions," Proceedings of the Conference on Matrix Structural Analysis, October, 1965.

G. C. BEST, "Derivation of Element Stiffness Matrices," *AIAA Journal* **2**, 8 (August 1964).

Y. K. CHEUNG and O. C. ZIENKIEWICZ, "Plates and Tanks on Elastic Foundation—An Application of Finite Element Method," *International Journal of Solids and Structures* **1** (1965).

Y. K. CHEUNG and J. O. PEDRO, "Automatic Preparation of Data Cards for the Finite Element Method," Research Report, Civil Engineering Division, School of Engineering, University College of Swansea, March 1966.

R. W. CLOUGH, "The Finite Element Method in Plane Stress Analysis," Proceedings of the 2nd ASCE Conference on Electronic Computation, September 1960.

R. W. CLOUGH and E. L. WILSON, "Stress Analysis of a Gravity Dam by the Finite Element Method" Report No. SESM-63-2, Institute of Engineering Research, University of California, Berkeley, Calif.

R. W. CLOUGH and E. L. WILSON, "Stress Analysis of a Gravity Dam by the Finite Element Method" Report No. SESM-63-2, Institute of Engineering Research, University of California, Berkeley, Calif. 1963.

R. W. CLOUGH and J. L. TOCHER, "Analysis of Thin Arch Dams by the Finite Element Method," Proceedings of the International Symposium on the Theory of Arch Dams, Southampton University, England, April 1964.

R. W. CLOUGH and YUSEF RASHID, "Finite Element Analysis of Axi-Symmetric Solids," ASCE Engineering Mechanics Division, February 1965.

R. W. CLOUGH and J. L. TOCHER, "Finite Element Stiffness Matrices for Analysis of Plate Bending," AFFDL-TR-66-80, Proceedings of the Conference on Matrix Methods of Structural Mechanics, Wright-Patterson Air Force Base, Ohio, October 1965.

CARL J. COSTANTINO, "Finite Element Approach to Stress Wave Problems," *ASCE Journal Engineering Mechanics Division*, April 1967.

D. J. DAWE, "A Finite Element Approach to Plate Vibration Problems," *Journal AIAA* **3** (1965).

D. J. DAWE, "Parallelogram Element in the Solution of Rhombic Cantilever Plate Problems," *Journal of Strain Analysis* **3** (1966).

J. ERGATOUDIS, "Quadrilateral Elements in Plane Analysis," M.Sc. Thesis, University of Wales, Swansea, 1966.

R. H. GALLAGHER, and J. PADLOG, "Discrete Element Approach to Structural Instability Analysis," *AIAA Journal* **1** 6 (1963).

R. J. GUYAN, "Distributed Mass Matrix for Plate Elements in Bending," *AIAA Journal* **3** (1965).

LEONARD R. HERMANN, "Finite-Element Bending Analysis for Plates," *ASCE Journal, Engineering Mechanics Division* **93** (October 1967).

GARDNER W. HICKS, "Finite-Element Elastic Buckling Analysis," *ASCE Journal, Structure Division* **93** (December 1967).

B. M. IRONS, "Numerical Integration Applied to Finite Element Method," Conference on Use of Digital Computers in Structural Engineering, University of Newcastle, July 1966.

R. E. JONES and D. R. STROME, "Direct Stiffness Method of Analysis of Shells of Revolution Utilizing Curved Element" *Journal AIAA* 1966.

KANWAR K. KAPUR and BILLY J. HARTZ, "Stability of Plates Using the Finite Element Method," *ASCE Journal Engineering Mechanics Division,* April 1966.

Z. A. LU, J. PENZIEN, and E. P. POPOV, "Finite Element of Solution for Thin Shells of Revolution," I.E.R. Technical Report SESM 63-3, University of California, Berkeley, Calif., September 1963.

T. H. H. PIAN, "Derivation of Element Stiffness Matrices," *AIAA Journal* **2,** 3 (March 1962).

T. H. H. PIAN, "Derivation of Element Stiffness Matrices by Assumed Stress Distributions," *AIAA Journal* **2** (July 1964).

T. H. H. PIAN, "Element Stiffness Matrices for Boundary Compatibility and for Prescribed Boundary Stresses," Proceeding Conference on Matrix Methods in Structure Mechanics, Air Force Institute of Technology, Wright Patterson Air Force Base, October 1966.

EGAR P. POPOV, JOSEPH PENZIEN, and ZUNG-AN LU, "Finite Element Solution for Axisymmetrical Shells," *ASCE Journal, Engineering Mechanics Division* (October 1964).

Y. RASHID, "Solution of Elasto-Static Boundary Value Problems by the Finite Element Method," Ph.D. thesis presented to the University of California, Berkeley, Calif., 1964.

R. T. SEVERN and D. R. TAYLOR, "The Finite Element Method for Flexure of Slabs when Stress Distributions are Assumed," *Proceedings of the Institute of Civil Engineering* **34** (1966).

T. L. TOCHER, "Analysis of Plate Bending Using Triangular Elements," Ph.D. dissertation, University of California, Berkeley, Calif., 1962.

JAMES L. TOCHER and BILLY J. HARTZ, "Higher-Order Finite Element for Plane Stress," ASCE Engineering Mechanics Division, August 1967.

OKTAY URAL, "Analysis of Totally Submerged Thin Shells by Finite Element Technique," Proceedings of International Association of Shell Structures, Pacific Symposium, October 1971.

OKTAY URAL, "Finite Element Method—A Versatile Tool for Civil Engineers," Proceedings of McGill-EIC Conference on Finite Element Method, Montreal, June 1972.

S. UTKU and F. A. AKYUZ, "ELAS—A General Purpose Computer Program for the Equilibrium Problems of Linear Structures," Jet Propulsion Laboratory, Pasadena, Calif., 1968.

W. VISSER, "A Finite Element Method for the Determination of Non-Stationary Temperature Distribution and Thermal Deformations," Proceedings of the Conference on Matrix Methods in Structural Mechanics, Air Force Institute of Technology, Dayton, Ohio, October 1965.

E. L. WILSON, "Finite Element Analysis of Two-Dimensional Structures," Report No. 63-2 of the University of California, Berkeley, Calif., June 1963.

O. C. ZIENKIEWICZ and Y. K. CHEUNG, "The Finite Element Method for Analysis of Elastic Isotropic and Orthotropic Slabs," *Proceedings of the Institute of Engineers,* London **28** (August 1964).

O. C. ZIENKIEWICZ and Y. K. CHEUNG, "Finite Element Method of Analysis for Arch Dam Shells, and Comparison with Finite Difference Procedure," Proceedings, International Symposium in the Theory of Arch Dams, Southampton University, England, April 1964.

O. C. ZIENKIEWICZ and Y. K. CHEUNG, "Finite Elements in the Solution of Field Problems," *The Engineer* **24** (September 1965).

O. C. ZIENKIEWICZ, B. IRONS, and B. NATH, "Natural Frequencies of Complex, Free or Submerged, Structures by the Finite Element Method," in *Symposium on Vibrations in Civil Engineering,* London (April 1965).

Index

A

Adaptability, 73
Adjoint method, 29
Akyuz, F.A., 69, 78
Argyris, J.H., 68, 77
Axial deformation, 47
Axisymmetric analysis, 143
    evaluation of, 152
    formulation of, 150
    stiffness matrix, 150
Axisymmetric solids, 141

B

Bell, E.T., 31

C

Carpenter, S.T., 65
Cayley, A., 2
Cholesky's method, 22, 29, 191
Clough, R.W., 68, 77, 98, 153
Compatibility condition, 34
Computer program, 63, 187
    BOSOR-3, 194
    comparative study, 195
    ELAS, 191
    FEPLS, 195
    general purpose, 186
    ICES-STRUDL, 188
    NASTRAN, 189
    SAAS-II, 190
    SLADE, 195
    special purpose, 186
    STRATA, 192
    Wilson, 194
    Zienkiewicz-Cheung, 193
Convergence, 72
Crout, P.D., 22, 31

D

Davis, R.L., 202
Degree of freedom, 80, 144

INDEX

Determinant
  co-factor expansion, 6
  definition, 6
Digital computer, 67
Displacement behavior, 155
Displacement continuity, 155
Displacement finite element method, 70
Displacement function
  comments, 177
Displacement matrix, 76, 93
Displacement method, 40
Dynamic program, 206

E

Elasticity matrix, 87, 161
Element
  constant strain triangle, 80
  triangular, 79
Element stiffness matrix, 94, 110
  generation, 112, 126
  superposition of, 94
Energy principle, 71, 90
  internal strain energy, 90
  potential energy, 90
Equilibrium equations, 34
Equivalent load matrix, 76

F

Finite difference method, 68
Finite element, 68
Finite element method, 40, 69, 206
  adaptability, 73
  additional applications, 257
  analysis of a gravity dam, 257
  analysis of a simple beam, 124
  applications, 237
  beam analysis, 124
  computer programs, 185
  convergence, 72
  definition, 68
  element, 71
  energy principle, 71
  flow through a channel, 262
  formulation, 70, 74, 100
  formulation of problems, 238
  general discussion, 101
  ICES-STRUDL, 205
  mesh generation, 74
  node, 68
  plane truss analysis, 100
  plate bending analysis, 222, 255
  plate with hole, 239
  plate in tension, 109, 218
  rigid frame analysis, 250

Finite element method (*cont.*)
  slope-stability analysis, 259
  steps for solution, 70
  stiffness matrix, 70
  thin plate analysis, 109
  trigonometric relations, 103, 104
  tunnel analysis, 260
Finite elements
  comparison, 181
  constant-strain triangle, 155
  plate bending element, 170
  rectangular plane stress element, 156
  rectangular plate bending element, 171
  triangular element with six nodes, 164
  triangular plate bending element, 177
  triangular torus, 143
Flexibility matrix, 37
Flexibility matrix method, 34, 45
Flow through a channel, 262
Force finite element method, 69
Force method, 35
Fortran IV, 63

G

Gauss-Jordan method, 12, 29
  operations, 13
General displacement function, 75, 81
Gere, S.M., 65
Gravity dam analysis, 257
Grinter, L.E., 68, 77

H

Hooke's law, 34
Hrennikoff, A., 68, 77

I

ICES, 206
  types of analysis, 207
ICES-STRUDL
  commands, 208
  element types, 212
  finite element method, 212
  plane stress elements, 80
  plate bending elements, 213
  scope, 207
  shell element, 213
Input matrices, 51
Inversion
  adjoint method, 9
  Cholesky method, 22
  Gauss-Jordan, 12
  partitioning, 16
Irons, B.M., 31

Isotropic plate, 175

K

Keith, D., 202
Kinematic determinacy, 35
Kinematically determinate system, 41
Kinematically indeterminate system, 42

L

Laursen, H.I., 65
Linear structural system, 33

M

Martin, H.C., 65
Matrix
  addition, 5
  adjoint, 8
  bandwidth, 28
  column, 3
  conformally partitioned, 17
  definition, 2
  determinant of, 4, 6
  diagonal, 4
  element of, 2
  equality, 4
  identity, 3
  inversion, 9
  method of analysis, 33
  multiplication, 5
  null, 3
  orthogonal, 4
  position, 85
  row, 3
  singular, 4, 27
  skew symmetric, 3
  square, 3
  subtraction, 5
  symmetric, 3, 25
  transpose of, 4
  triangular, 4
Maxwell-Betti law, 37, 44
McHenry, D., 68, 77
Meek, J.L., 69, 78
Melosh, R.J., 31, 69, 77
Mesh generation, 74, 197, 198, 199, 214, 218, 223, 230, 240, 250, 253, 259, 260, 261
  coarse, 74

N

Nodal displacements, 76
Nodal points, 80

Node numbering, 95, 96

P

Partitioning method, 29
Pian, T.H.H., 69, 77
Plane
  strain, 75, 80
  stress, 76, 79
Plane strain analysis, 79
Plane stress development, 79
  theoretical development, 80
Positive set, 28
Principal stresses, 76, 201
Przemiemecki, J.S., 65

R

Rashid, Y., 69, 77, 153
Rubenstein, M.F., 65

S

Slope stability analysis, 259
Statically determinate system, 36
Statically indeterminate system, 38
Stiffness matrix, 44, 92, 93
  element, 94
  properties, 44
  superposition, 94, 106
  total, 95, 96
Stiffness matrix method, 34, 40, 45
  degree of kinematic indeterminacy, 63
  input quantities, 63
  output quantities, 63
Strain
  initial, 87
  temperature, 87
Strains, 75, 87
Stress, 76, 87, 200
  principal, 76
Surface traction, 91
System of linear equations, 2, 82

T

Tapered beam, 196
Theory of elasticity, 34
Thin plate theory
  basic assumptions, 170
  finite element theory, 170
Timoshenko, S., 78, 98
Tocher, J.L., 69, 77

Toug, P., 69, 77
Tunnel analysis, 260
Turner, J.J., 98

U

Ural, O., 65, 69, 78
Utku, S., 69, 78

W

Wang, C.K., 65, 153
Weaver, W., 65
Wilson, E.L., 69, 77, 153

Z

Zienkiewicz, O.C., 69, 78